Mathematics for the IB Diploma
Higher Level Topic 8
Statistics and Probability

Hugh Neill and Douglas Quadling

Series Editor Hugh Neill

13.12

CAMBRIDGE
UNIVERSITY PRESS

CAMBRIDGE UNIVERSITY PRESS
Cambridge, New York, Melbourne, Madrid, Cape Town, Singapore, São Paulo, Delhi

Cambridge University Press
The Edinburgh Building, Cambridge CB2 8RU, UK

www.cambridge.org
Information on this title: www.cambridge.org/9780521714631

First published 2008

A catalogue record for this publication is available from the British Library

ISBN   978-0-521-71463-1 paperback

The authors and the publishers are grateful to the following examination boards for permission to reproduce
questions from past examination papers, identified in the text as follows.
OCR   Oxford, Cambridge and RSA Examinations
IBO   International Baccalaureate Organization
The authors, and not the examination boards, are responsible for the method and accuracy of the answers
to examination questions given; these may not necessarily constitute the only possible solutions.

This material has been developed independently of the International Baccalaureate Organization (IBO).
This text is in no way connected with, nor endorsed by, the IBO.

Contents

Introduction

Statistics and Probability has been written especially for the International Baccalaureate Mathematics HL and FM examinations. This book covers the syllabus for Topic 8.

It is assumed that students will already have completed the statistics chapters in the Higher Level Books 1 and 2. There is a small amount of material which extends a topic beyond the syllabus as printed, with the aim of enhancing students' appreciation of the subject. This is indicated by an asterisk (*) at the appropriate place in the text.

Occasionally within the text paragraphs appear in *this type style*. These paragraphs are usually outside the main stream of the mathematical argument, but may help to give insight, or suggest extra work or different approaches.

Students are expected to have access to graphic display calculators and the text places considerable emphasis on their potential for supporting the learning of statistics.

Numerical work is presented in a form intended to discourage premature approximation. In ongoing calculations inexact numbers appear in decimal form like 3.456... , signifying that the number is held in a calculator to more places than are given. Numbers are not rounded at this stage; the full display could be, for example, 3.456 123 or 3.456 789. Final answers are then stated with some indication that they are approximate, for example '1.23 correct to 3 significant figures'.

There are plenty of exercises. At the end of the book there is a Review exercise which includes some questions from past International Baccalaureate examinations, but on a different syllabus. At the time of writing, there are few current versions of the Higher Level examinations, so there is no backlog of examination questions on the newer parts of the syllabus.

The author thanks OCR and IBO for permission to use some examination questions and Cambridge University Press for their help in producing this book. Particular thanks are due to Jane Miller for permission to use her material from the Cambridge Advanced Level Mathematics series and to Sharon Dunkley and Linda Moore for their help and advice. However, the responsibility for the text, and for any errors, remains with the author.

1 Cumulative distribution functions

A cumulative distribution function is an alternative way of describing the probability distribution of a random variable. When you have completed this chapter, you should

- know the meaning of a cumulative distribution function and how to find it for both continuous and discrete random variables
- be able to find the probability density from a cumulative distribution function.

1.1 Finding the cumulative distribution from the probability density

When you have statistical data about a continuous random variable, you can represent it either with a histogram or with a cumulative frequency diagram. A similar choice can be made for graphs representing theoretical probability models.

One possibility is to use a probability density function $f(x)$. This is conventionally defined over the complete set $\mathbb{R}$ of real numbers, although often there are intervals $]-\infty, a]$ or $[b, \infty[$ for which $f(x) = 0$.

Probability density functions have the following properties (see Higher Level Book 2 Section 10.2).

- $f(x) \geq 0$ for all $x \in \mathbb{R}$.

- The area under the probability density graph is equal to 1. That is,

$$\int_{-\infty}^{\infty} f(x)\, dx = 1.$$

- The probability that the random variable X lies in the interval $[c, d]$ is equal to the area under the probability density graph over this interval. That is,

$$P(c \leq X \leq d) = \int_{c}^{d} f(x)\, dx.$$

Another possibility is to use a **cumulative distribution function** $F(x)$, which is defined as the probability that the random variable X is less than or equal to x. That is,

$$F(x) = P(X \leq x).$$

For a continuous random variable it is not important whether you define the function as $P(X \leq x)$ or $P(X < x)$, since the probability that X is equal to any particular single value x is 0. But the distinction is important if you extend the definition to include discrete random variables.

Example 1.1.1

Fig. 1.1 shows the graph of a probability density function

$$f(x) = \begin{cases} 1 - \frac{1}{2}x & \text{for } 0 \le x \le 2, \\ 0 & \text{otherwise.} \end{cases}$$

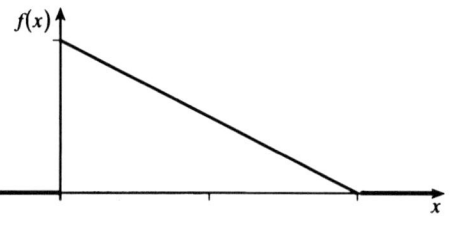

Fig. 1.1

Find the cumulative distribution function $F(x)$, and draw its graph.

When x is negative the probability density is 0, so $F(x) = 0$ for $x \le 0$. For $0 \le x \le 2$, the cumulative probability is equal to the area under the line in the interval $[0, x]$, which is shown by the shaded region in Fig. 1.2. This is a trapezium with parallel sides of lengths 1 and $1 - \frac{1}{2}x$, at a distance x apart. Therefore

$$F(x) = \tfrac{1}{2}\left(1 + \left(1 - \tfrac{1}{2}x\right)\right) \times x = x - \tfrac{1}{4}x^2.$$

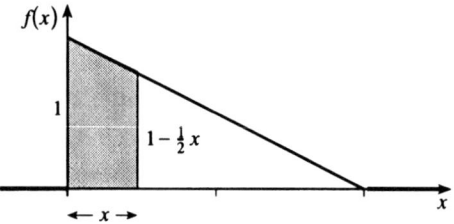

Fig. 1.2

This gives $F(2) = 2 - \frac{1}{4} \times 2^2 = 2 - 1 = 1$; you can check that this is the area of the triangle. Beyond $x = 2$ the probability density is again 0, so that there is no further increase in the cumulative probability. Therefore $F(x) = 1$ for $x > 2$.

The graph of $F(x)$ is shown in Fig. 1.3.

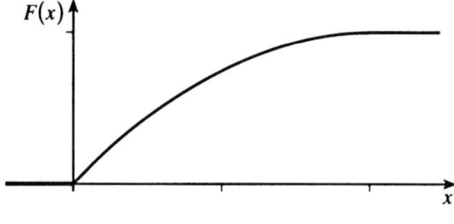

Fig. 1.3

In most cases finding $F(x)$ for a given $f(x)$ involves integration, but this presents a small problem of notation. If you write the probability $P(X \le x)$ as $\int f(x)\,dx$ evaluated over the interval $\left]-\infty, x\right]$, then the letter x appears twice, once inside the integral and the other as one of the limits of integration. Fortunately, in finding a *definite* integral it doesn't matter what letter you use inside the integral: the value of $\int_c^d f(x)\,dx$ is exactly the same as $\int_c^d f(t)\,dt$ or $\int_c^d f(u)\,du$. So it is better to avoid the problem by using a different letter inside the integral, writing the cumulative probability for example as

$$P(X \le x) = \int_{-\infty}^x f(t)\,dt.$$

If a continuous random variable has probability density $f(x)$, the cumulative distribution function $F(x) = P(X \le x)$ is given by

$$\int_{-\infty}^x f(t)\,dt.$$

Example 1.1.2

A probability density function is defined by the equation

$$f(x) = \begin{cases} 0 & \text{for } x < 1, \\ \dfrac{1}{x^2} & \text{for } x \geq 1. \end{cases}$$

(a) Check that this is a valid probability density function.

(b) Find an expression for the cumulative distribution function $F(x)$.

(a) It is obvious that $f(x) \geq 0$ for all x, so you just have to prove that $\displaystyle\int_{-\infty}^{\infty} f(x)\,dx = 1$.

$$\int_{-\infty}^{\infty} f(x)\,dx = \int_{-\infty}^{1} 0\,dx + \int_{1}^{\infty} \frac{1}{x^2}\,dx.$$

The first integral on the right is clearly 0, and the second is the limit as $v \to \infty$ of

$$\int_{1}^{v} \frac{1}{x^2}\,dx = \left[-\frac{1}{x}\right]_{1}^{v} = 1 - \frac{1}{v}.$$

Since $\displaystyle\lim_{v\to\infty}\left(1 - \frac{1}{v}\right) = 1 - 0 = 1$, it follows that

$$\int_{-\infty}^{\infty} f(x)\,dx = 0 + 1 = 1.$$

(b) If $x < 1$, then clearly $F(x) = 0$.
If $x \geq 1$ then

$$F(x) = P(X \leq x)$$

$$= \int_{-\infty}^{1} 0\,dt + \int_{1}^{x} \frac{1}{t^2}\,dt$$

$$= 0 + \left[-\frac{1}{t}\right]_{1}^{x}$$

$$= 1 - \frac{1}{x}.$$

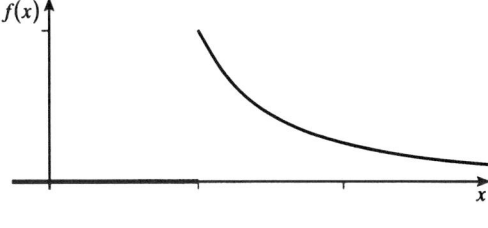

Fig. 1.4

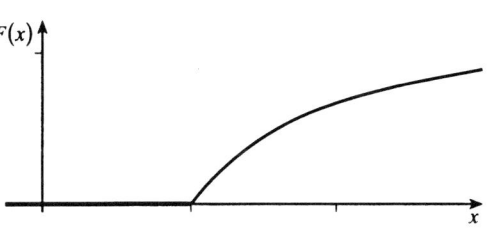

Fig. 1.5

The graph of $f(x)$ is shown in Fig. 1.4, and that of $F(x)$ in Fig. 1.5.

Sometimes to describe $f(x)$ you need different non-zero expressions over different intervals of the domain. This is illustrated in the next example.

Example 1.1.3
In a population of a particular breed of dog the age distribution is modelled by the probability density function shown in Fig. 1.6, with equation

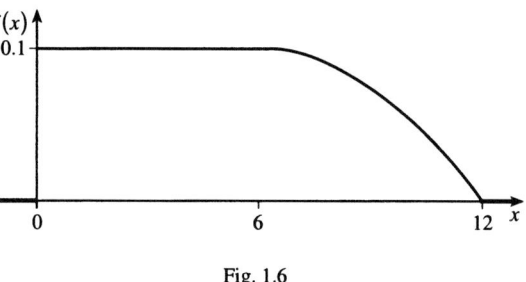

Fig. 1.6

$$f(x) = \begin{cases} \frac{1}{10} & \text{for } 0 < x \le 6, \\ \frac{1}{30}x - \frac{1}{360}x^2 & \text{for } 6 < x \le 12, \\ 0 & \text{otherwise,} \end{cases}$$

where x denotes the age in years. Find expressions for the cumulative distribution function. Hence find the median and quartile ages of the dogs in the population.

Clearly $F(x) = 0$ for $x \le 0$. For $0 < x \le 6$ the region under the probability density graph is a rectangle with width x and height $\frac{1}{10}$, so that $F(x) = \frac{1}{10}x$. For $6 < x \le 12$ the region under the graph is a rectangle of width 6 and height $\frac{1}{10}$ and a region under a curve (see the shaded area in Fig. 1.7) so

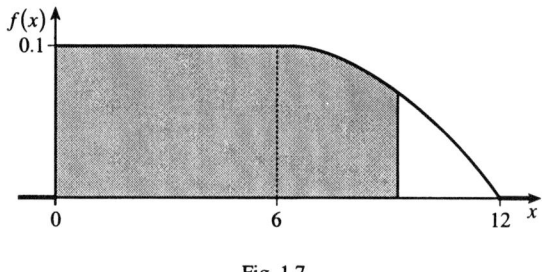

Fig. 1.7

$$F(x) = \frac{6}{10} + \int_6^x \left(\frac{1}{30}t - \frac{1}{360}t^2\right) dt$$

$$= \frac{6}{10} + \left[\frac{1}{60}t^2 - \frac{1}{1080}t^3\right]_6^x$$

$$= \frac{6}{10} + \left(\frac{1}{60}x^2 - \frac{1}{1080}x^3\right) - \left(\frac{36}{60} - \frac{216}{1080}\right)$$

$$= \frac{1}{1080}\left(216 + 18x^2 - x^3\right).$$

For $x > 12$ the probability density is 0, so the value of $F(x)$ remains constant and equal to

$$F(12) = \frac{1}{1080}(216 + 2592 - 1728) = \frac{1}{1080} \times 1080 = 1.$$

Putting all this together,

$$F(x) = \begin{cases} 0 & \text{for } x \le 0, \\ \frac{1}{10}x & \text{for } 0 < x \le 6, \\ \frac{1}{1080}\left(216 + 18x^2 - x^3\right) & \text{for } 6 < x \le 12, \\ 1 & \text{for } x > 12. \end{cases}$$

The graph of $F(x)$ is shown in Fig. 1.8.

From Higher Level Book 2 Section 10.4 the median M is the value of x such that $P(X \le M) = \frac{1}{2}$, that is $F(M) = \frac{1}{2}$. Since $F(6) = \frac{6}{10} > \frac{1}{2}$, M lies in the interval $]0,6]$, so that $F(M) = \frac{1}{10}M$. Therefore $\frac{1}{10}M = \frac{1}{2}$, so $M = 5$.

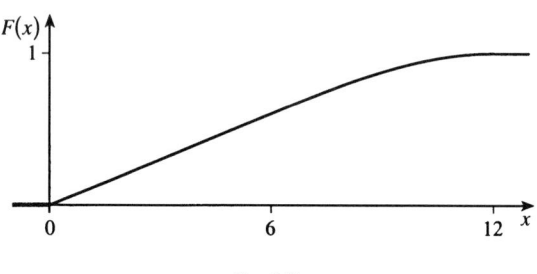

Fig. 1.8

Similarly the lower quartile Q_1 lies in the interval $]0,6]$. Therefore $\frac{1}{10}Q_1 = \frac{1}{4}$, so $Q_1 = 2.5$.

However, the upper quartile Q_3 lies in the interval $]6,12]$, so that

$$F(Q_3) = \tfrac{1}{1080}\left(216 + 18Q_3{}^2 - Q_3{}^3\right).$$

Since $F(Q_3)$ has to equal $\frac{3}{4}$, $x = Q_3$ has to satisfy the equation

$$\tfrac{1}{1080}\left(216 + 18x^2 - x^3\right) = \tfrac{3}{4}, \quad \text{which is} \quad x^3 - 18x^2 + 594 = 0.$$

Using a calculator to solve this equation, the root in the interval $]6,12]$ is $7.533\ldots$.

So the upper quartile $Q_3 = 7.53$, correct to 3 significant figures.

1.2 Finding the probability density from the cumulative distribution

Since the cumulative distribution is found from the probability density by integrating, you can find the probability density from the cumulative distribution by differentiating. For example, in Example 1.1.1 the cumulative distribution function over the interval $[0,2]$ is $F(x) = x - \frac{1}{4}x^2$. Differentiating this gives $F'(x) = 1 - \frac{1}{2}x$, which is the equation for the line segment joining $(0,1)$ to $(2,0)$ in the probability density graph.

> If a continuous random variable has cumulative distribution function $F(x)$, the probability density function $f(x)$ is given by
>
> $$f(x) = F'(x).$$

There is one small complication in applying this rule. Although the graph of $F(x)$ is always a continuous line, it may contain isolated points where there is a sudden change of direction, such as $(0,0)$ in Fig. 1.3 and $(1,0)$ in Fig. 1.5. You can't differentiate $F(x)$ at these points, so the equation $f(x) = F'(x)$ doesn't apply there. This doesn't matter, since $f(x)$ is only used to calculate probabilities over an interval, not at isolated points. The usual convention is to define $f(x)$ at these points so that the intervals used to describe $f(x)$ are the same as those used to describe $F(x)$.

In some applications it is easier to begin by finding the cumulative distribution and then to find the probability density by differentiation.

Example 1.2.1

A bad darts player is equally likely to hit any point on the board. (He is allowed to ignore any throws which miss the board completely.) The radius of the board is a. Find the cumulative distribution and the probability density for the random variable R, the distance from the centre at which a dart hits the board.

The distance of any hit from the centre must be between 0 and a, so the probability density is zero in the intervals $r < 0$ and $r > a$.

If r is between 0 and a, the hits for which $R \le r$ lie inside (or on the circumference of) a circle of radius r, shown shaded in Fig. 1.9.

Since the hits are uniformly distributed over the surface of the board, the probability that $R \leq r$ is equal to the ratio of the area of this circle to the area of the whole board. That is,

$$F(r) = P(R \leq r) = \frac{\pi r^2}{\pi a^2} = \frac{r^2}{a^2}.$$

So, over the interval $0 \leq r < a$, the probability density is given by

$$f(r) = F'(r) = \frac{2r}{a^2}.$$

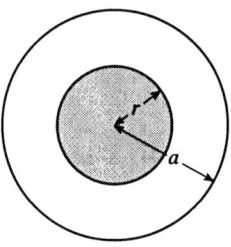

Fig. 1.9

The equations for the cumulative distribution and the probability density are then

$$F(r) = \begin{cases} 0 & \text{for } r < 0, \\ \dfrac{r^2}{a^2} & \text{for } 0 \leq r \leq a, \\ 1 & \text{for } r > a \end{cases} \quad \text{and} \quad f(r) = \begin{cases} 0 & \text{for } r < 0, \\ \dfrac{2r}{a^2} & \text{for } 0 \leq r \leq a, \\ 0 & \text{for } r > a. \end{cases}$$

1.3   Extension to discrete probability distributions

The definition of the cumulative distribution as $P(X \leq x)$ is still valid if the random variable X is discrete, but the value of this probability remains constant in the intervals between successive elements of the sample space. The graph of the cumulative distribution function therefore consists of a set of line segments parallel to the x-axis.

Example 1.3.1
Draw graphs of the probability distribution and of the cumulative distribution function for the Poisson probability $\text{Po}(2.5)$.

Your calculator probably has a program to calculate cumulative Poisson probabilities, and you should make sure that you know how to use this. However, you may only be able to use it to find the probability that $X \leq n$ for a particular value of n, and to draw the graph you want a table of values of $P(X \leq n)$ for $n = 0, 1, 2, 3, \ldots$. So it may be more convenient to generate these as a sequence from the Poisson formula

$$P(X = n) = e^{-m} \frac{m^n}{n!}$$

with $m = 2.5$.

This is very simple, because the factor e^{-m} doesn't involve n, and

$$\frac{m^n}{n!} = \frac{m \times m^{n-1}}{n \times (n-1)!} = \frac{m}{n} \times \frac{m^{n-1}}{(n-1)!}.$$

So, if you use u_n to denote the probability $P(X = n)$, values of u_n can be found from the inductive definition

$$u_0 = e^{-m}, \qquad u_n = \frac{m}{n} \times u_{n-1} \text{ for } n = 1, 2, 3, \ldots .$$

Now in this example you want not only values of $P(X = n)$ but also those of $P(X \le n)$, which is the sum sequence of u_n (see Higher Level Book 1 Section 30.4). So if you denote $P(X \le n)$ by v_n, values of v_n can be found from the inductive definition

$$v_0 = u_0 = e^{-m}, \qquad v_n = v_{n-1} + u_n = v_{n-1} + \frac{m}{n} \times u_{n-1} \text{ for } n = 1, 2, 3, \ldots .$$

With $m = 2.5$, these equations give the values (to 3 significant figures) for u_n and v_n in Table 1.10.

n	0	1	2	3	4	5	6	...
$u_n = P(X = n)$	0.082	0.205	0.257	0.214	0.134	0.067	0.028	...
$v_n = P(X \le n)$	0.082	0.287	0.544	0.758	0.891	0.958	0.986	...

Table 1.10

To draw the cumulative distribution graph, you need to define $F(x) = P(X \le x)$ not just when x is a natural number 0, 1, 2, ... but when x is any real number. Using the values of $P(X \le n)$ given in Table 1.10, it follows that

$$F(x) = \begin{cases} 0 & \text{for } x < 0, \\ 0.082 & \text{for } 0 \le x < 1, \\ 0.287 & \text{for } 1 \le x < 2, \\ 0.544 & \text{for } 2 \le x < 3, \end{cases} \qquad \text{and so on.}$$

The graph of the probability distribution u_n is given in Fig. 1.11, and that of the cumulative distribution is given in Fig. 1.12. In the latter graph, a small dot has been added at the left end of each line segment to make clear that the jump occurs immediately *before* the integer value of x.

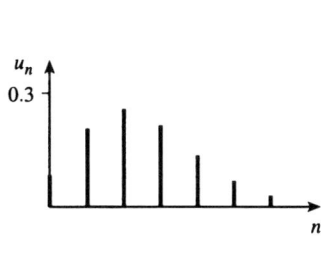

Fig. 1.11

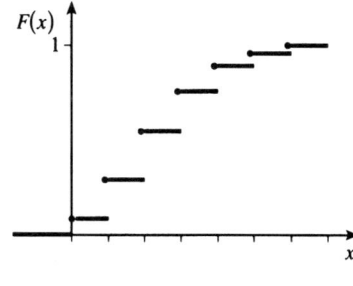

Fig. 1.12

Exercise 1

1   The probability density function of a random variable, X, is

$$f(x) = \begin{cases} \frac{1}{8}x & 0 \le x \le 4, \\ 0 & \text{otherwise.} \end{cases}$$

(a) Find the cumulative distribution function of X.   (b) Find the median.

2   The probability density function of a continuous random variable, X, is

$$f(x) = \begin{cases} \frac{2}{3} & 0 \le x < 1, \\ \frac{4}{3} - \frac{2}{3}x & 1 \le x \le 2, \\ 0 & \text{otherwise.} \end{cases}$$

(a) Find $F(x)$.   (b) Find the lower quartile.   (c) Find the 85th percentile.

3   The probability density function of a continuous random variable, X, is

$$f(x) = \begin{cases} \frac{3}{14}x & 0 \le x < 2, \\ \frac{3}{28}x(4 - x) & 2 \le x \le 4, \\ 0 & \text{otherwise.} \end{cases}$$

(a) Find $F(x)$.   (b) Find the median.   (c) Find $P(2.5 \le X \le 3)$.

4   Daily sales of petrol, X, at a service station (in tens of thousands of litres) are distributed with probability density function

$$f(x) = \begin{cases} \frac{3}{4}x(2 - x)^2 & 0 \le x \le 2, \\ 0 & \text{otherwise.} \end{cases}$$

(a) Find $F(x)$.   (b) Find the median in litres.

5   The cumulative distribution of a random variable, X, is

$$F(x) = \begin{cases} 0 & x < 0, \\ \frac{1}{108}\left(9x^2 - x^3\right) & 0 \le x \le 6, \\ 1 & x > 6. \end{cases}$$

Find the probability density function of X.

6   The cumulative distribution function of a random variable, X, is

$$F(x) = \begin{cases} 0 & x < 0, \\ \frac{1}{18}x^2 & 0 \le x < 3, \\ \frac{2}{3}x - \frac{1}{18}x^2 - 1 & 3 \le x \le 6, \\ 1 & x > 6. \end{cases}$$

Find the probability density function of X.

7   The random variable X metres represents the side of a square. The area of the square is represented by the random variable Y metre^2.

(a) If X has uniform probability density 0.2 over the interval $]0,5[$, find the cumulative distribution function $F(x) = P(X \le x)$.

(b) Deduce the cumulative distribution function for the area, $G(y) = P(Y \le y)$.

(c) Hence find the probability density function $g(y)$ for the area of the square.

8 The random variable X metres represents the base of a rectangle of area 1 metre^2. The height is represented by the random variable Y metres.

    (a) If X has uniform probability density 1 over the interval $]0,1[$, find the cumulative distribution function $F(x) = \mathrm{P}(X \le x)$.

    (b) Deduce the cumulative distribution function for the height, $G(y) = \mathrm{P}(Y \le y)$.

    (c) Hence find the probability density function $g(y)$ for the height of the rectangle.

9* In Example 1.1.3, suppose that the probability density function for the age distribution of the dog population remains the same over time, and that the total number of dogs in the population remains constant. It can then be shown that, if the random variable Y represents the age in years to which a dog lives, then the probability density function for Y is

$$g(y) = -yf'(y),$$

where f is the probability density function defined in Example 1.1.3. Use this to find the median age to which a dog of this breed lives.

10 A random variable X has binomial probability $\mathrm{B}(3,0.4)$. Find the values of the cumulative distribution function in the intervals $]-\infty,0[$, $[0,1[$, $[1,2[$, $[2,3[$ and $[3,\infty[$.

11 Draw the graph of the cumulative distribution function for Poisson probability $\mathrm{Po}(1.7)$.

2 Geometric and exponential probability

The two probability models described in this chapter both arise from the same question: how long must you wait until a certain event occurs? When you have completed this chapter, you should

- be able to calculate and apply geometric and exponential probability
- know expressions for the expectation and variance of these probability models
- understand the connection between exponential and Poisson probability.

2.1 The geometric distribution

Consider the following three examples.

1   A dice is rolled until a six is scored. Let X be the number of rolls up to and including the roll on which the first six occurs.

2   A card is selected, with replacement, from a standard pack of cards until an ace is drawn. Let Y be the number of selections up to and including the one on which the first ace occurs.

3   A person has a 1 in 17 chance of winning a prize in a lottery. She keeps playing the lottery once each week. Let W be the number of weeks up to and including the week in which she first wins a prize.

These three random variables have certain similarities.

In the first example,

$$P(X = 1) = P(\text{a six occurs on the first throw}) = \tfrac{1}{6}.$$

Let s represent the event that a six occurs on a trial and let f represent the event that a six does not occur. Then $X = 2$ corresponds to the sequence fs. Similarly the event $X = 3$ corresponds to the sequence ffs. Notice that there is only one possible sequence for each value of X.

You can now calculate the probability distribution for X:

$$P(X = 2) = P(fs) = \left(\tfrac{5}{6}\right) \times \left(\tfrac{1}{6}\right) = \tfrac{5}{36},$$

$$P(X = 3) = P(ffs) = \left(\tfrac{5}{6}\right)^2 \times \tfrac{1}{6} = \tfrac{25}{216},$$

$$P(X = 4) = P(fffs) = \left(\tfrac{5}{6}\right)^3 \times \tfrac{1}{6} = \tfrac{125}{1296},$$

and so on.

You can generalise this as

$$P(X = x) = \left(\tfrac{5}{6}\right)^{x-1} \times \tfrac{1}{6} \quad \text{for } x = 1, 2, 3, \dots \ .$$

In the second example the probability distribution formula for Y will be $P(Y = y) = \left(\tfrac{12}{13}\right)^{y-1} \times \tfrac{1}{13}$ for values of $y = 1, 2, 3, \dots$ since

$$P(Y = y) = P\left(\overbrace{ff\dots f}^{y-1}s\right) = \overbrace{\tfrac{12}{13} \times \tfrac{12}{13} \times \dots \times \tfrac{12}{13}}^{y-1 \text{ of these}} \times \tfrac{1}{13}.$$

In the third example, by a similar argument,

$$P(W = w) = \left(\tfrac{16}{17}\right)^{w-1} \times \tfrac{1}{17} \quad \text{for } w = 1, 2, 3, \dots \ .$$

You will know from experience that when you roll a dice you may have to wait a long time before a six turns up. In fact, there is no upper limit to the possible values of X, Y and W for these distributions. The sample space is the complete set $\mathbb{Z}^+$ of positive integers.

In general, if you have a sequence of trials for which the probabilities of success and failure are p and q (where $q = 1 - p$), and the random variable X is the number of trials up to and including the first success, then

$$P(X = x) = q^{x-1} \times p.$$

The first few probabilities for this distribution are set down in Table 2.1.

x	1	2	3	4	$\dots$
$P(X = x)$	p	qp	$q^2 p$	$q^3 p$	$\dots$

Table 2.1

You will see that these probabilities form a geometric sequence with first term p and common ratio q (see Higher Level Book 1 Section 30.1). For this reason the probability distribution for X is called a **geometric distribution**. Notice that the values of the probabilities depend on just one parameter p (since q is $1 - p$). A shorthand method of stating that the random variable X has this distribution is to write $X \sim \text{Geo}(p)$.

Fig. 2.2 shows graphs of the distributions $\text{Geo}(0.4)$ and $\text{Geo}(0.6)$. Notice that, because a larger value of p gives a smaller value of q, the graph tails off more rapidly when p is larger. But in both cases the mode of the distribution is 1; before you start to roll the dice, you are more likely to get the first six on the first roll than on any subsequent roll.

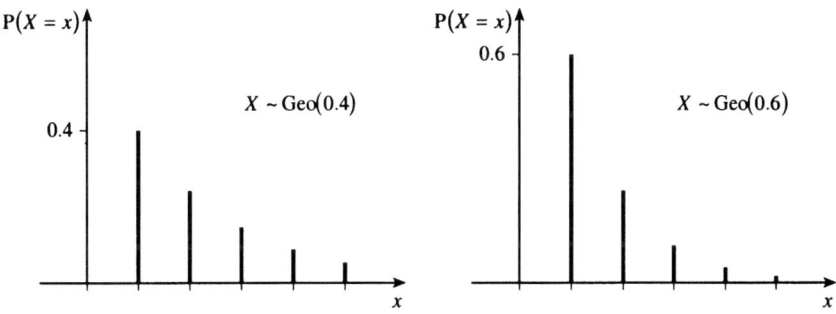

Fig. 2.2

The conditions which have to be satisfied for a set of trials to produce a geometric distribution are the same as those for a binomial distribution (see Higher Level Book 1 Section 34.2). But now, instead of asking how many successes you get in a fixed number of trials, the question is how many trials you must perform until you get a success.

The geometric distribution

- A single trial has exactly two possible outcomes (success and failure) and these are mutually exclusive.
- The outcome of each trial is independent of the outcome of all the other trials.
- The probability of success at each trial is constant.
- The trials are repeated until a success occurs.

The random variable X, which represents the number of trials up to and including the first success, then has a probability distribution given by the formula

$$P(X = x) = pq^{x-1} \quad \text{for } x = 1, 2, 3, \dots ,$$

where p is the probability of success, and $q = 1 - p$ is the probability of failure.

When the random variable X satisfies these conditions $X \sim \text{Geo}(p)$.

Example 2.1.1

A darts player must start a game by getting a dart in a certain region of the dart-board, The probability that any single dart hits this region is $\frac{2}{5}$. The player keeps throwing darts until he hits the required region. Calculate the probability that he needs

(a) exactly 5 throws,

(b) at most 2 throws,

(c) at least 8 throws,

(d) at most 10 throws.

(e) How many throws does the player need to make to be at least 99.999% sure of hitting the required region?

(f) What assumptions have you made in using the geometric distribution to model this situation?

Let X be the number of throws needed to hit the required region of the dartboard. X has a geometric distribution, with parameter $p = \frac{2}{5}$.

(a)   You need to find $P(X = 5)$. From the probability distribution formula

$$P(X = 5) = \frac{2}{5} \times \left(\frac{3}{5}\right)^4 = 0.0518, \text{ correct to 3 significant figures.}$$

(b)   You need to find $P(X \le 2)$. This can be found by adding probabilities.

$$P(X \le 2) = P(X = 1) + P(X = 2)$$
$$= \tfrac{2}{5} + \tfrac{3}{5} \times \tfrac{2}{5}$$
$$= 0.64.$$

(c)   Finding $P(X \ge 8)$ is best done by realising that, if the player needs 8 or more throws, the first 7 must have been failures Therefore

$$P(X \ge 8) = \left(\tfrac{3}{5}\right)^7 = 0.0280, \text{ correct to 3 significant figures.}$$

(d)   You can find $P(X \le 10)$ from the probability of the complementary event, by using $P(X \le 10) = 1 - P(X \ge 11)$. Then, using a similar argument to that of part (c),

$$P(X \ge 11) = \left(\tfrac{3}{5}\right)^{10}$$

so

$$P(X \le 10) = 1 - P(X \ge 11)$$
$$= 1 - \left(\tfrac{3}{5}\right)^{10} = 0.994, \text{ correct to 3 significant figures.}$$

(e)   To find m such that $P(X \le m) \ge 0.999\,99$, use the argument in part (d) to get

$$P(X \le m) = 1 - P(X \ge m + 1) = 1 - \left(\tfrac{3}{5}\right)^m.$$

So, to find m, you must solve the inequality $1 - \left(\tfrac{3}{5}\right)^m \ge 0.999\,99$, which is the same as $\left(\tfrac{3}{5}\right)^m \le 0.000\,01$, or $\left(\tfrac{5}{3}\right)^m \ge 100\,000$.

Inequalities like this can be solved by taking logarithms of both sides (see Higher Level Book 1 Section 31.5). Using logarithms to base 10,

$$\log\left(\tfrac{5}{3}\right)^m \ge \log 100\,000$$
$$m \log\left(\tfrac{5}{3}\right) \ge 5$$
$$m \ge \frac{5}{\log \tfrac{5}{3}} = 22.53\ldots \,.$$

The smallest integer satisfying this is 23, so this is the required number of throws.

(f)   The assumptions which have been made are that the probability of success, here $\tfrac{2}{5}$, is constant for every throw. This means that the darts player does not improve with practice, nor does he get worse as he gets tired. The results of individual throws are also assumed to be independent, so that each throw has no effect on any other throw.

You will have noticed that in Example 2.1.1 different methods were used to find $P(X \le 2)$ and $P(X \le 10)$. It is interesting to compare the two methods by using them to find $P(X \le m)$ for a general geometric distribution $\text{Geo}(p)$.

Method 1   This uses the formula for the sum of a geometric series.

$$P(X \le m) = P(X = 1) + P(X = 2) + P(X = 3) + \ldots + P(X = m)$$
$$= p + pq + pq^2 + \ldots + pq^{m-1}$$
$$= p\left(1 + q + q^2 + \ldots + q^{m-1}\right)$$
$$= p\frac{1 - q^m}{1 - q}$$
$$= p\frac{1 - q^m}{p}$$
$$= 1 - q^m.$$

Method 2   Begin by noting that $P(X \le m) = 1 - P(X \ge m + 1)$, and that if $m + 1$ or more trials are needed then the first m trials must have resulted in failure. So

$$P(X \ge m + 1) = q^m,$$

and $P(X \le m) = 1 - q^m$.

For the geometric distribution $X \sim \text{Geo}(p)$

$$P(X \ge x) = (1 - p)^{x-1}.$$

The cumulative geometric distribution function is

$$P(X \le x) = 1 - (1 - p)^x.$$

2.2   Expectation and variance of a geometric distribution

When you find a new probability distribution it is a good idea to check that the sum of the probabilities for all possible values of the random variable is equal to 1. For geometric probability this sum is an infinite series

$$\sum_{x=1}^{\infty} P(X = x) = p + pq + pq^2 + \ldots$$
$$= p\left(1 + q + q^2 + \ldots\right).$$

The expression in brackets is an infinite geometric series with sum $\frac{1}{1 - q}$, so

$$\sum_{x=1}^{\infty} P(X = x) = p \times \frac{1}{1 - q}$$
$$= p \times \frac{1}{p} = 1,$$

as expected.

Your calculator may have programs to find geometric and cumulative geometric probabilities. You could use them to answer questions like those in Example 2.1.1 parts (a) to (d).

The expressions for the expectation and the variance are also found as infinite series. A method of calculating these is given in the Appendix Section A.2.

The results are:

For the geometric distribution $\text{Geo}(p)$,

$$E(X) = \mu = \frac{1}{p} \quad \text{and} \quad \text{Var}(X) = \sigma^2 = \frac{q}{p^2},$$

where $q = 1 - p$.

As a check, use these expressions to find the expectation and variance in the examples at the beginning of Section 2.1. For example, the distribution for rolling a dice is $\text{Geo}\left(\frac{1}{6}\right)$, for which $E(X) = \frac{1}{\frac{1}{6}} = 6$ and

$\text{Var}(X) = \dfrac{\frac{5}{6}}{\left(\frac{1}{6}\right)^2} = 30$. Is it reasonable that on average the number of rolls needed for a six to come up is 6?

Try carrying out the experiment a lot of times (or simulate it with a computer), and calculate the mean and variance of your results. How well do they agree with the theoretical values?

Example 2.2.1

A brand of cereal has a card in each packet. The cards form a set of 50 different pictures. They are distributed at random in the packets. A child has collected 49 of the picture cards and so needs one more to complete the set. What is the mean number of packets she will need to open to obtain the last picture?

The probability that the next packet she opens will contain the picture she needs is $\frac{1}{50} = 0.02$. So X, the number of packets opened, up to and including the first one to contain the required picture card, has a geometric distribution with parameter $p = 0.02$.

So $E(X) = \dfrac{1}{p} = \dfrac{1}{0.02} = 50$.

The child can expect to open 50 packets to obtain the last picture.

Example 2.2.2

The random variable Y has a geometric distribution. If $P(Y = 2) = 0.24$ and $P(Y = 3) = 0.144$, find the expected value and the variance of Y.

If the geometric distribution has parameter p, then

$$P(Y = 2) = qp = 0.24, \quad \text{and} \quad P(Y = 3) = q^2 p = 0.144.$$

Therefore $\dfrac{q^2 p}{qp} = q = \dfrac{0.144}{0.24} = 0.6$.

Thus $p = 0.4$ and $\text{E}(Y) = \dfrac{1}{0.4} = 2.5$, $\text{Var}(Y) = \dfrac{0.6}{0.4^2} = 3.75$.

Example 2.2.3

Two children play a game on the beach. They begin by drawing a circle in the sand. Then they each move back 5 metres and pick up a pebble. They throw their pebbles together and try to get them into the circle. Ann's probability of success is 0.25, and Ben's is 0.2. If they throw together again and again, what is the expectation of the number of turns needed until

(a) they both succeed on the same turn,   (b) one or other of them succeeds?

(a)   The probability that they will both succeed on the same turn is $0.25 \times 0.2 = 0.05$, and the probability that they won't both succeed is 0.95. So the probability that they first both succeed on the xth turn is $0.95^{x-1} \times 0.05$. The distribution is therefore $\text{Geo}(0.05)$, with expectation
$$\frac{1}{0.05} = 20.$$

(b)   The probabilities that one of them, Ann or Ben, will fail on any turn are 0.75 and 0.8 respectively. The probability that they will both fail is therefore $0.75 \times 0.8 = 0.6$. If one or other of them (or both) succeeds, then they don't both fail, so the probability is $1 - 0.6 = 0.4$.
The distribution is therefore $\text{Geo}(0.4)$, with expectation $\dfrac{1}{0.4} = 2.5$.

Exercise 2A

In this exercise give probabilities to 4 decimal places.

1   The random variable X represents the number of trials up to and including a success. The probability of a success on any given trial is $p = 0.3$, independently of any other trial. Find the probability that

(a) $X = 5$,   (b) $X = 6$,   (c) $X = 5, 6$ or 7.

2   The random variable Y represents the number of trials up to and including a success. The probability of a success on any given trial is $p = 0.4$, independently of any other trial. Find the probability that

(a) $Y = 3$,   (b) $Y = 4$,   (c) $Y = 2, 3$ or 4.

3   Trials are repeated until a success is obtained. The probability of a success on any given trial is 0.8, independently of any other trial. Find the probability that the total number of trials needed, including the successful one, is

(a) exactly 3,   (b) less than or equal to 3,   (c) more than 5.

4   The random variable G has a geometric distribution, and the probability of a success on the first trial is 0.025. Find the probability that the first success

(a) occurs on the 10th trial,

(b) occurs after the 10th trial,

(c) occurs on or before the 10th trial.

5 It is given that the random variable T, which can take values 1, 2, 3, ... , has a geometric distribution and $P(T = 1) = 0.15$. Calculate

(a) $P(T > 7)$,   (b) $P(T > 11)$,   (c) $P(T \geq 12)$,   (d) $P(T \leq 11)$.

6 A card is chosen at random from an ordinary pack, and replaced after its value has been noted. The process continues until a picture card (K, Q, J) is obtained. Z is the total number of cards drawn, including the picture card. Find

(a) $P(Z = 5)$,   (b) $P(Z > 7)$,   (c) $P(Z \leq 7)$.

7 In order to start a board game each player must throw at least one six with a pair of dice. Find the probability that for Gaye to start, she needs

(a) one throw,   (b) five throws,   (c) more than eight throws.

8 Wayne is counting cars going past the front gate of the school. He has been told that, on average, one car in 12 on the roads is green. Assuming that car colours are independent of each other, find the probability that

(a) the first green car he sees is the 8th that passes,

(b) the first green car is not among the first 15 cars,

(c) there is at least one green car among the first 10 cars.

9 A certain irrational number has a decimal expansion in which each digit is randomly chosen from the set $\{0, 1, 2, 3, 4, 5, 6, 7, 8, 9\}$. Find the probability that

(a) the first 9 occurs in the 8th place,   (b) there is no 9 in the first 10 places.

10 It is known that, on average, one person in three at a shopping centre is wearing trainers. A market researcher observes people entering the shopping centre. Use a geometric distribution, with $p = \frac{1}{3}$, to calculate the probability that

(a) the first person wearing trainers is the fourth person observed,

(b) the first person wearing trainers is not among the first five people observed,

(c) the first person wearing trainers is either the fourth or the fifth person observed.

Give a reason why a geometric distribution might not be suitable in this context.

11 It is known that 9% of the population belongs to blood group B. A hospital consultant visits patients in a hospital and determines their blood groups, stopping when he has found one patient of blood group B.

(a) What is the probability that he needs to examine at least 11 patients?

(b) How many patients must he examine to be 99.8% confident of finding at least one in blood group B?

12 A biologist is collecting data about fruit flies (*Drosophila*). She knows that one fruit fly in 10 has a striped body, and this property occurs at random in the population of fruit flies. She needs to collect one fruit fly with a striped body.

(a) Find the probability that she needs to collect at least 15 fruit flies.

(b) Find the number of fruit flies she must collect to be 99.99% sure of obtaining one with a striped body.

13   An ordinary cubical dice is thrown repeatedly. A six is obtained on the 5th throw. Find the probability that the next six is

   (a) obtained on the 12th throw,   (b) not obtained until at least the 13th throw.

14   The random variable S can take values 1, 2, 3, … and has a geometric distribution. It is given that $P(S = 2) = 0.2244$. Find the value of $P(S = 1)$, given that it is less than 0.5.

15   The random variable X has a geometric distribution with $P(X > 1) = 0.75$. Find $E(X)$ and $Var(X)$.

16   A fair coin is tossed until a head is obtained.

   (a) Find the expected value and the variance of the number of tosses required.

   (b) What is the expected value of the number of tails tossed?

17   A box of Chocobix cereal costs $1.50. One box in 25, on average, contains a silver button. If I buy a box of this cereal every week, find the expected total cost of the boxes bought, up to and including the first box with a silver button.

2.3   The exponential distribution

Geometric probability arises when you have a sequence of trials, which you go on performing until a certain event occurs. Exponential probability is similar but, instead of a sequence of separate trials, the event may occur at any instant in a continuous stream of time. Here are some examples.

1 A web site receives on average 15 hits per hour. How long will it be before the next hit?

2 A radioactive substance emits on average 5 particles per second. When you switch on the counter, how long will it be before it registers the first emission?

3 On average a car drives down a village street once every 2 minutes. If a villager opens her front door, how long will it be before the next car passes?

It is not difficult to guess what form the probability density function might take in examples like these. You met a similar situation in the chapter on exponential decay (Higher Level Book 1 Section 32.2) where the equation $f(t) = ab^t$ could be used either with domain $\mathbb{N}$ if values were observed discretely, or with domain $\mathbb{R}^+$ if the decay occurred continuously through time. And it was shown later that, for continuous decay, it is often more convenient to write b as e^k, so that the equation becomes $f(t) = ae^{kt}$ (see Higher Level Book 2 Section 13.3).

You can get from geometric to exponential probability in a similar way. In place of the expression pq^{x-1} for the probability that X takes the value x, you have a probability density function for a continuous random variable X of the form

$$f(x) = ce^{-\lambda x} \quad \text{for } x > 0.$$

The negative sign is inserted so that λ is a positive constant. (Remember that for exponential decay b is between 0 and 1, so that k is negative. Here, similarly, the probability q is between 0 and 1, so the constant in the exponent is negative.)

The next step is to find the constant c. Since f is a probability density, the area under the probability density graph must be 1. So c must be chosen so that

$$\int_0^\infty f(x)\, dx = 1.$$

Now

$$\int_0^\infty e^{-\lambda x}\, dx = \lim_{v \to \infty} \int_0^v e^{-\lambda x} dx$$

$$= \lim_{v \to \infty} \left[-\frac{1}{\lambda} e^{-\lambda x} \right]_0^v$$

$$= \lim_{v \to \infty} \left(-\frac{1}{\lambda} e^{-\lambda v} + \frac{1}{\lambda} \right)$$

$$= 0 + \frac{1}{\lambda} = \frac{1}{\lambda}.$$

So $c \times \dfrac{1}{\lambda}$ must equal 1, that is $c = \lambda$. The equation of exponential probability density for $x > 0$ is therefore

$$f(x) = \lambda e^{-\lambda x}.$$

This is called the **exponential probability distribution.** It is denoted by $\mathrm{Exp}(\lambda)$. Like geometric probability, there is only one parameter λ. This is the average number of occurrences of the event in unit time. For example, in the first example above, x is measured in hours and $\lambda = 15$, so the probability density function is $f(x) = 15e^{-15x}$.

You will already have noticed that exponential probability deals with just the same situations as Poisson probability. But the questions asked about them are different. Poisson probability is concerned with how many events occur in a unit of time; exponential probability is about the time that you have to wait until the next event occurs. But the conditions for the probability model to be valid are the same in both cases.

The exponential distribution

This distribution applies to events which

- occur randomly in space or time
- occur singly, that is they cannot occur simultaneously
- occur independently
- occur at a constant rate.

The random variable X, which represents the distance or time up to the first occurrence of the event, then has a probability density function

$$f(x) = \begin{cases} \lambda e^{-\lambda x} & \text{for } x \geq 0, \\ 0 & \text{otherwise,} \end{cases}$$

where λ is the mean rate at which the event occurs.

When the random variable satisfies these conditions, $X \sim \mathrm{Exp}(\lambda)$.

Often it is more convenient to use the cumulative distribution function than the probability density function. For $x \geq 0$, this is given by

$$F(x) = P(X \leq x) = \int_0^x \lambda e^{-\lambda t} \, dt$$

$$= \left[-e^{-\lambda t} \right]_0^x$$

$$= 1 - e^{-\lambda x}.$$

If $X \sim \text{Exp}(\lambda)$, the cumulative exponential distribution function for X is

$$F(x) = \begin{cases} 1 - e^{-\lambda x} & \text{for } x \geq 0, \\ 0 & \text{otherwise.} \end{cases}$$

The graph of the probability density function is shown in Fig. 2.3, and that of the cumulative distribution function in Fig. 2.4, both for $\lambda = 1.5$.

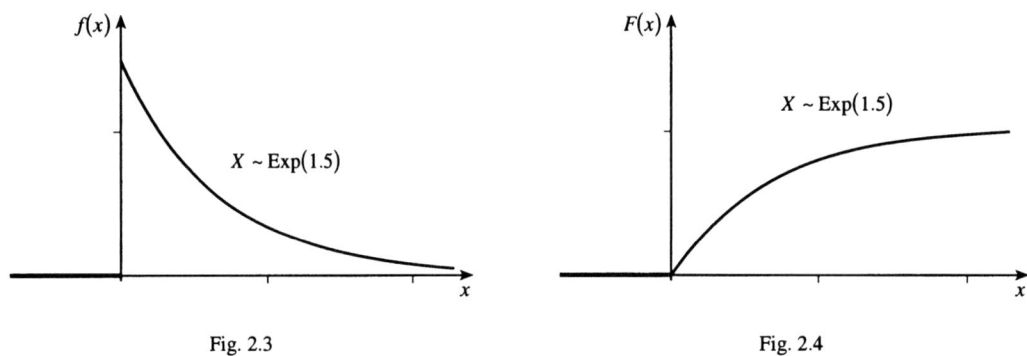

Fig. 2.3   Fig. 2.4

Example 2.3.1
In a packing plant apples are transported along a conveyor belt. On average a checker spots a substandard apple once every 15 seconds. What is the probability that she will miss at least one if her attention is diverted for 6 seconds?

You must begin by choosing what unit of time to use. If you choose a minute, then the mean rate of appearance of substandard apples is 4 per minute, so the appropriate probability distribution is $\text{Exp}(4)$.

The random variable X is the number of minutes until the next substandard apple appears after the checker's attention wanders, and you want to know the probability that this is less than 0.1. From the expression for the cumulative probability,

$$P(X \leq 0.1) = 1 - e^{-4 \times 0.1} = 1 - e^{-0.4} = 0.330, \text{ correct to 3 significant figures.}$$

The probability that the checker will miss at least one substandard apple is 0.330.

(If you had used a second rather than a minute as the unit of time, then λ would equal $\frac{1}{15}$ and you would calculate $P(X \leq 6)$. You can easily check that this gives the same value for the probability.)

2.4   Expectation and variance of an exponential distribution

Calculating the mean and variance for the exponential distribution requires the calculation of infinite integrals (see the Appendix Section A.6.) The results are:

> For the exponential distribution $X \sim \text{Exp}(\lambda)$,
>
> $$E(X) = \mu = \frac{1}{\lambda} \quad \text{and} \quad \text{Var}(X) = \sigma^2 = \frac{1}{\lambda^2}.$$

Example 2.4.1

A traffic census finds that on average a van passes a checkpoint once every 4 minutes. A new person comes on duty. Find the mean and standard deviation of the time he should expect to wait until the next van passes.

Taking a minute as the unit of time, the frequency of vans passing per minute is $\frac{1}{4}$. So the waiting time up to the next van, X minutes, has distribution $\text{Exp}\left(\frac{1}{4}\right)$. This has $\mu = \frac{1}{\frac{1}{4}} = 4$ and

$\sigma^2 = \frac{1}{\left(\frac{1}{4}\right)^2} = 16$, so that $\sigma = 4$.

The expectation is 4 minutes and the standard deviation is 4 minutes.

2.5   Combining two exponential distributions

Suppose that a taxi firm has two telephone lines, and that on average the numbers of calls received per hour on these lines are a and b. What is the probability distribution for the time that will elapse before a call comes in on either line?

You will probably guess that it is $\text{Exp}(a+b)$, as it would be if all the calls came in on one line. This is in fact correct, but it is not quite obvious.

Denote by A and B the random variables for the number of hours before the first call comes in on the two lines, and let X be the random variable for all the calls taken together. Then $A \sim \text{Exp}(a)$ and $B \sim \text{Exp}(b)$.

It is easiest to concentrate on the probability that a call has *not* come in. Suppose that after x hours no call has come in to the office. In that case no call has come in on either line. So

$$X > x \quad \Leftrightarrow \quad (A > x) \cap (B > x).$$

Now suppose that the flows of calls on the two lines are independent of each other. (You would have to assume that each call is so short that there is no risk of a customer finding the first line engaged and switching to the second.) Then it follows that

$$P(X > x) = P(A > x) \times P(B > x).$$

And since $A \sim \text{Exp}(a)$ and $B \sim \text{Exp}(b)$, you know that

$$P(A > x) = e^{-ax} \quad \text{and} \quad P(B > x) = e^{-bx},$$

so

$$P(X > x) = e^{-ax} \times e^{-bx} = e^{-(a+b)x},$$

and the cumulative distribution for X is

$$F(x) = P(X \le x) = 1 - e^{-(a+b)x}.$$

So X has probability distribution $\text{Exp}(a+b)$.

What this means is that, if you have two random sequences of events running in parallel over the same period of time which satisfy the conditions for exponential probability, you can combine them into a single random sequence satisfying the same conditions.

> If two independent sequences of events are taking place in parallel, such that the times to the next occurrence are distributed as $\text{Exp}(a)$ and $\text{Exp}(b)$ respectively, then the time to the next occurrence of one or other event is distributed as $\text{Exp}(a+b)$.

2.6* Proof of the rule for exponential probability

In Section 2.3 the expression for the probability density, $f(x) = \lambda e^{-\lambda x}$, was justified by analogy with exponential decay. This section shows how to get it by a probability argument.

Because the random variable is continuous, the proof uses differentiation. It is simplest to begin by finding the cumulative distribution function, or rather its complement, the probability that the event has *not* occurred up to the time x. If this is denoted by y, then

$$y = P(X > x).$$

The argument is explained in terms of time, because most applications are to events which take place in continuous time. Exponential probability can also be applied to the spacing of points along a line, in which case the wording of the argument would need to be changed, but the essential ideas remain the same.

Now suppose that x is increased to a value $x + \delta x$. Then the value of y will change to $y + \delta y$, and the problem is to find an expression for δy in terms of δx. The basic idea is:

If (1) the event has not occurred up to time $x + \delta x$, then
(2) it has not occurred up to time x, and
(3) it does not occur between times x and $x + \delta x$.

You know the probabilities of (1) and (2): they are $P(X > x + \delta x) = y + \delta y$ and $P(X > x) = y$. To make up an equation, you also need an expression for the probability of (3).

Suppose that δx is a very short interval of time. There is a possibility that an event could occur in this time, but the possibility that more than one event could occur is negligible. (Remember that one of the conditions is that events cannot occur simultaneously.) Also, the probability that an event occurs in this time is proportional to the duration of the interval, δx. So it can be denoted by $\lambda \times \delta x$, where λ is a constant.

The probability of (3), that an event does not occur during this interval is therefore $1 - \lambda \delta x$. Also events occur independently, so (3) is independent of (2). Therefore

$$y + \delta y = y \times (1 - \lambda \delta x),$$

which can be written more simply as

$$\delta y = -y\lambda \delta x, \quad \text{or} \quad \frac{\delta y}{\delta x} = -\lambda y.$$

Although this is written as an exact equation, it is more properly an approximation, because there is still the remote possibility that two events might be fitted into the interval δx. But this becomes more and more improbable as δx gets smaller. So in the limit, as $\delta x \to 0$,

$$\frac{dy}{dx} = -\lambda y.$$

This is a standard differential equation for y in terms of x. It is proved in Higher Level Book 2 Section 23.3 that its solution has the form

$$y = Ce^{-\lambda x}$$

where C is constant. And since $y = P(X > x)$, and you know that $P(X > 0) = 1$, the constant C must be equal to 1. That is,

$$P(X > x) = e^{-\lambda x}.$$

So the cumulative distribution function $F(x) = P(X \leq x)$ is

$$F(x) = 1 - e^{-\lambda x}.$$

Finally, the probability density function $f(x) = F'(x)$ is

$$f(x) = \lambda e^{-\lambda x}.$$

Exercise 2B

1   If $X \sim \text{Exp}(0.2)$, find
   (a) $P(X \leq 1)$,   (b) $P(X > 10)$,   (c) $P(4 \leq X \leq 6)$.

2   On average 5 cyclists pass the window each minute. What is the probability that, when you look out of the window, a cyclist will pass in the next 15 seconds?

3   On average a robin visits a garden bird table once every half-hour. Assuming these visits occur randomly, what is the probability that a robin will visit the bird table within the next 5 minutes?

4   An ambulance unit gets an emergency call once every 10 minutes on average. What is the probability that no calls will come in between 10.15 and 10.30?

5   A woman with a cold sneezes randomly 10 times an hour on average. Find the probability that she will not sneeze in the next 5 minutes.

6   Suppose that a major earthquake occurs somewhere in the world once every 4 years on average. What is the probability that a major earthquake will occur next year?

7   On a long street some of the paving stones are damaged. On average there are 25 damaged paving stones per kilometre. Assuming that these are randomly distributed, and that there are no gaps due to side streets, what is the probability that there are no damaged paving stones within 50 metres on either side of the post office entrance?

8   For the distribution $X \sim \text{Exp}(\lambda)$
   (a) find (i) the median, (ii) the quartiles, (iii) the mode;
   (b) what percentile is represented by the mean?
   Show your answers on a sketch of the probability density function for $\text{Exp}(1)$.

9   If X has an exponential distribution, and if $\text{Exp}(X) = 10$, find $P(5 \le x \le 15)$.

10   If $X \sim \text{Exp}(\lambda)$, and $P(X \le 4) = 0.2$, find $\text{E}(X)$.

11   If X has an exponential distribution, find the probability that X lies within one standard deviation of the mean.

12   If $X \sim \text{Exp}(\lambda)$ and $P(1 < X < 2) = 0.24$, find two possible values for λ.

13   On average a car alarm goes off accidentally in the neighbourhood once every 2 hours, and a house alarm once every 3 hours. What is the expected time before the next alarm (of either kind) goes off?

14   At a city bus stop buses seem to arrive at random. The average waiting time for a number 10 bus is 10 minutes, and the average waiting time for a number 15 bus is 15 minutes.
   (a) How long should a person expect to have to wait before a bus comes along?
   (b) What is the probability that the person will have to wait more than 20 minutes?

3 Linear combinations of random variables

This chapter is about the distribution of linear functions and linear combinations of random variables. When you have completed it, you should

- know how to calculate the mean and variance of $aX + b$ when the mean and variance of X are known
- know how to calculate the mean and variance of $aX + bY$ when the mean and variance of X and Y are known.

3.1 Linear functions of a single random variable

Taxi fares are often calculated as the sum of two parts:

- a fixed charge, which appears on the meter as soon as you get in the cab
- an amount proportional to the distance travelled.

So for a journey of x kilometres the total charge $\$y$ is found from an equation of the form $y = ax + b$, where $\$b$ is the fixed charge and a is the distance rate in dollars per kilometre.

Suppose that a taxi firm is reviewing the operation of its fleet of taxis. These will be used for journeys of many different distances, so a variety of fares will be charged. The distances and fares are both random variables, X kilometres and $\$Y$, and the firm could model its business by assigning probabilities to the various values of X. The same probabilities would also apply to the corresponding values of Y; that is, if a journey of x_i kilometres costs $\$y_i$, then $P(Y = y_i)$ is equal to $P(X = x_i)$. From these probabilities the firm could calculate the mean and variance of X and Y.

It is not difficult to guess how these are related. Since each separate value of X is multiplied by a to get the variable part of the charge, the mean will also be multiplied by a. Adding the fixed charge to each fare adds the same amount to the mean. So, if $Y = aX + b$,

$$E(Y) = aE(X) + b.$$

However, adding the fixed charge has no effect on the spread of values of Y, so the variance depends only on the rate per kilometre. And since the variance is calculated from the squares of values of the random variable,

$$\text{Var}(Y) = a^2 \text{Var}(X).$$

Another application of these rules is to situations where the same quantity is measured in two different systems of units, as in Example 3.1.1.

Example 3.1.1
The temperature in degrees Fahrenheit at a seaside resort is a random variable with mean 59 and variance 27. Find the mean and variance of the temperature in degrees Celsius.

Let X be the temperature in °F. Then $E(X) = 59$ and $\text{Var}(X) = 27$.

To convert a temperature from degrees Fahrenheit to degrees Celsius, you subtract 32 then multiply by $\frac{5}{9}$. Thus if $X\,°F$ is the same as $Y\,°C$, then $Y = \frac{5}{9}(X-32) = \frac{5}{9}X - \frac{160}{9}$.

So

$$\begin{aligned} E(Y) &= E\left(\tfrac{5}{9}X - \tfrac{160}{9}\right) \\ &= \tfrac{5}{9}E(X) - \tfrac{160}{9} \\ &= \tfrac{5}{9} \times 59 - \tfrac{160}{9} = 15, \end{aligned}$$

and

$$\begin{aligned} \operatorname{Var}(Y) &= \left(\tfrac{5}{9}\right)^2 \operatorname{Var}(X) \\ &= \tfrac{25}{81} \times 27 = \tfrac{25}{3}. \end{aligned}$$

The proofs that $E(Y) = aE(X) + b$ and $\operatorname{Var}(Y) = a^2 \operatorname{Var}(X)$ are simple if X is a discrete random variable. Suppose that X takes the value x_i with probability p_i. The essential part of the argument is that the values of $y_i = ax_i + b$ are all different, so that each y_i also has probability p_i.

There is a trivial exception to this statement, if $a = 0$. Then all the values of Y are equal to b, so that $E(Y) = b$ and $\operatorname{Var}(Y) = 0$. Obviously the results are true in this case.

The expected value of Y is given by

$$\begin{aligned} E(Y) &= \mu_Y \\ &= \sum y_i p_i \\ &= \sum (ax_i + b) p_i \\ &= a\sum x_i p_i + b\sum p_i \\ &= aE(X) + b \qquad \left(\text{since } \sum p_i = 1\right). \end{aligned}$$

Since $Y = aX + b$, $E(Y) = E(aX + b)$; it follows that $E(aX + b) = aE(X) + b$.

Also

$$\begin{aligned} \operatorname{Var}(Y) &= \sum (y_i - \mu_Y)^2 p_i \\ &= \sum ((ax_i + b) - (a\mu_X + b))^2 p_i \qquad (\text{since } \mu_Y = a\mu_X + b) \\ &= \sum (a(x_i - \mu_X))^2 p_i \\ &= \sum a^2 (x_i - \mu_X)^2 p_i \\ &= a^2 \sum (x_i - \mu_X)^2 p_i \\ &= a^2 \operatorname{Var}(X). \end{aligned}$$

Since $Y = aX + b$, $\operatorname{Var}(Y) = \operatorname{Var}(aX + b)$; it follows that $\operatorname{Var}(aX + b) = a^2 \operatorname{Var}(X)$.

These results are still true if X is a continuous random variable. The proof is essentially the same, but it is more difficult to write in terms of integrals and probability density.

For any random variable X,

$$E(aX + b) = aE(X) + b,$$

$$\operatorname{Var}(aX + b) = a^2 \operatorname{Var}(X),$$

where a and b are constants.

3.2 Linear combinations of more than one random variable

Situations often arise in which you know the expected value and variance of each of several random variables and need to find the expected value and variance of a linear combination of these. For example, you know the expected value and variance of the thickness of the sheets which make up a laminated windscreen and want to find the expected value and variance of the total thickness of the windscreen.

In order to investigate any possible relations between expected values and variances you could start by considering two discrete random variables, X and Y. The question is, if you know the mean and variance of X and Y, how do you find the mean and variance of the random variable $Z = X + Y$?

Example 3.2.1 does this numerically.

Example 3.2.1

Members of a club compete in pairs for a cup. They perform as individuals, and the two scores are then added together. Before the competition one pair reckons the probability distributions for their scores, X and Y, to be as shown in Tables 3.1 and 3.2.

X	50	60	70
Probability	0.2	0.7	0.1

Table 3.1

Y	30	40	50
Probability	0.1	0.3	0.6

Table 3.2

(a) Find the probability distribution for their joint score $X + Y$.

(b) Find the expectation and variance for the individual scores and for the joint score.

(a) The joint score can take the values 80, 90, 100, 110 or 120.

The probabilities of these are calculated as follows.

$$P(X + Y = 80) = P(X = 50 \cap Y = 30)$$
$$= P(X = 50) \times P(Y = 30)$$
$$= 0.2 \times 0.1 = 0.02,$$
$$P(X + Y = 90) = P(X = 50 \cap Y = 40) + P(X = 60 \cap Y = 30)$$
$$= P(X = 50) \times P(Y = 40) + P(X = 60) \times P(Y = 30)$$
$$= 0.2 \times 0.3 + 0.7 \times 0.1 = 0.13, \text{ and so on.}$$

Continuing in this way gives the complete probability distribution shown in Table 3.3.

$X + Y$	80	90	100	110	120
Probability	0.02	0.13	0.34	0.45	0.06

Table 3.3

(b)   You can use a calculator to find the expectations and variances. Check for yourself that

$$E(X) = 59, \qquad E(Y) = 45, \qquad E(X + Y) = 104,$$

$$\text{Var}(X) = 29, \qquad \text{Var}(Y) = 45, \qquad \text{Var}(X + Y) = 74.$$

You will see that in this example

$$E(X + Y) = E(X) + E(Y) \qquad \text{and} \qquad \text{Var}(X + Y) = \text{Var}(X) + \text{Var}(Y).$$

To see if this is true in general you must carry out similar calculations algebraically.

To keep the algebra simple, suppose that X may take just three values x_1, x_2, x_3 and that Y may take just two values y_1, y_2. How many values can Z take?

Since any value of X can be combined with any value of Y, Z can take any of the values

$$x_1 + y_1, \quad x_2 + y_1, \quad x_3 + y_1, \quad x_1 + y_2, \quad x_2 + y_2, \quad x_3 + y_2.$$

A convenient way of showing this is with a two-way table like Table 3.4.

		Values of X			
		x_1	x_2	x_3	Total
Values of Y	y_1	r	s	t	q_1
	y_2	u	v	w	q_2
	Total	p_1	p_2	p_3	1

Table 3.4

The entries r, s, t, u, v, w in the table are the probabilities associated with each of the six values of Z; for example, $r = P(X = x_1 \cap Y = y_1) = P(Z = x_1 + y_1)$. These can be used to write the expectation of Z as

$$E(Z) = (x_1 + y_1)r + (x_2 + y_1)s + (x_3 + y_1)t + (x_1 + y_2)u + (x_2 + y_2)v + (x_3 + y_2)w.$$

It is possible that not all the six values of Z are different; for example, $x_1 + y_2$ might be equal to $x_3 + y_1$. This does not affect the mathematical argument which follows.

If you multiply out the brackets on the right and then rearrange the terms, you can write this as

$$E(Z) = x_1(r + u) + x_2(s + v) + x_3(t + w) + y_1(r + s + t) + y_2(u + v + w).$$

You will see that the first three brackets in this arrangement are the sums of the probabilities in the three columns of Table 3.1, and the last two brackets are the sums of the probabilities in the two rows. If you denote these sums by p_1, p_2, p_3 and q_1, q_2 as shown in Table 3.4, then they are just the probabilities associated with the various values of X and Y. For example,

$$p_1 = r + u = P(X = x_1 \cap Y = y_1) + P(X = x_1 \cap Y = y_2)$$

and

$$q_1 = r + s + t = P(X = x_1 \cap Y = y_1) + P(X = x_2 \cap Y = y_1) + P(X = x_3 \cap Y = y_1),$$

and these are just $P(X = x_1)$ and $P(Y = y_1)$ respectively. Using similar expressions for the other columns and rows in Table 3.4, the equation for $E(Z)$ becomes

$$E(Z) = (x_1 p_1 + x_2 p_2 + x_3 p_3) + (y_1 q_1 + y_2 q_2)$$
$$= E(X) + E(Y).$$

Notice that nothing has been said about the random variables. The result is true whether or not X and Y are independent.

> For any random variables X and Y,
>
> $$E(X + Y) = E(X) + E(Y).$$

This is probably what you would expect. But you may find the corresponding result for variance more surprising.

To start with, it helps to write $E(X)$ as μ_X, and similarly for Y and Z, and to write the variance in the form

$$\mathrm{Var}(X) = \sum_{i=1}^{3} x_i^2 p_i - \mu_X^2$$
$$= E(X^2) - \mu_X^2.$$

Similarly

$$\mathrm{Var}(Z) = E((X+Y)^2) - \mu_Z^2.$$

It has just been proved that $\mu_Z = \mu_X + \mu_Y$, so

$$\mathrm{Var}(Z) = E(X^2 + 2XY + Y^2) - (\mu_X^2 + 2\mu_X \mu_Y + \mu_Y^2).$$

By extending the result in the shaded box above to the sum of three random variables (remember that they need not be independent), you can write

$$E(X^2 + 2XY + Y^2) = E(X^2) + E(2XY) + E(Y^2),$$

and you have already seen in Section 3.1 that $E(2XY) = 2E(XY)$.

So, rearranging the terms,

$$\mathrm{Var}(Z) = (E(X^2) - \mu_X^2) + (E(Y^2) - \mu_Y^2) + 2(E(XY) - \mu_X \mu_Y)$$
$$= \mathrm{Var}(X) + \mathrm{Var}(Y) + 2(E(XY) - \mu_X \mu_Y).$$

Now look at the last bracket and go back to Table 3.4. You will see that

$$E(XY) = x_1y_1r + x_2y_1s + x_3y_1t + x_1y_2u + x_2y_2v + x_3y_2w,$$

and

$$\mu_x\mu_y = (x_1p_1 + x_2p_2 + x_3p_3)(y_1q_1 + y_2q_2)$$
$$= x_1y_1p_1q_1 + x_2y_1p_2q_1 + x_3y_1p_3q_1 + x_1y_2p_1q_2 + x_2y_2p_2q_2 + x_3y_2p_3q_2.$$

So $E(XY)$ and $\mu_x\mu_Y$ would be the same if

$$r = p_1q_1, \quad s = p_2q_1, \quad t = p_3q_1, \quad u = p_1q_2, \quad v = p_2q_2, \quad w = p_3q_2.$$

What do these equations mean? The first, for example, states that

$$P(X = x_1 \cap Y = y_1) = P(X = x_1) \times P(Y = y_1),$$

and the others are similar. These are just the conditions for the random variables X and Y to be independent. So, if X and Y are independent, then $E(XY) - \mu_X\mu_Y = 0$, and

$$Var(Z) = Var(X) + Var(Y).$$

You have probably found this proof rather daunting, but this is largely because of the complicated notation with all the suffixes – hence the decision to restrict X and Y to three and two values respectively. But obviously the proof would be essentially the same however many values there are. Try reading it through again, concentrating on the outline of the argument rather than the details. The result also holds if X and Y are continuous random variables, but it is more difficult to work with integrals rather than sums.

> If X and Y are independent random variables, then
> $$Var(X + Y) = Var(X) + Var(Y).$$

It is also worth noticing another useful result which has emerged during the proof.

> If X and Y are independent random variables, then
> $$E(XY) = E(X) \times E(Y).$$

Example 3.2.2

My journey to work is made up of four stages: a walk to the bus-stop, a wait for the bus, a bus journey and a walk at the other end. The times taken for these four stages are independent random variables U, V, W and X with expected values (in minutes) of 4.7, 5.6, 21.6 and 3.7 respectively and standard deviations of 1.1, 1.2, 3.1 and 0.8 respectively. What is the expected time and standard deviation for the total journey?

The expected time for the whole journey is

$$E(U + V + W + X) = E(U) + E(V) + E(W) + E(X)$$
$$= 4.7 + 5.6 + 21.6 + 3.7 = 35.6.$$

Since the variables U, V, W and X are independent,

$$\text{Var}(U + V + W + X) = \text{Var}(U) + \text{Var}(V) + \text{Var}(W) + \text{Var}(X)$$
$$= 1.1^2 + 1.2^2 + 3.1^2 + 0.8^2$$
$$= 12.9,$$

so the standard deviation for the whole journey is $\sqrt{12.9} = 3.59$, correct to 3 significant figures.

The relations between expected values and variances can be generalised to the situation where $Z = aX + bY$ and a and b are constants. By combining the results in this section with those in Section 3.1,

$$E(Z) = E(aX + bY)$$
$$= E(aX) + E(bY)$$
$$= aE(X) + bE(Y);$$

and for independent X and Y

$$\text{Var}(Z) = \text{Var}(aX + bY)$$
$$= \text{Var}(aX) + \text{Var}(bY)$$
$$= a^2\text{Var}(X) + b^2\text{Var}(Y).$$

> For any random variables X and Y, and constants a and b,
>
> $$E(aX + bY) = aE(X) + bE(Y).$$
>
> If X and Y are independent,
>
> $$\text{Var}(aX + bY) = a^2\text{Var}(X) + b^2\text{Var}(Y).$$

Example 3.2.3

The length, L (in cm), of the rectangular panels produced by a machine is a random variable with mean 26 and variance 4 and the width, B (in cm), is a random variable with mean 14 and variance 1. The variables L and B are independent. What are the expected value and variance of

(a) the perimeter of the panels,

(b) the difference between the length and the width?

(c)* What is the expected value of the area of the panels?

    (a)   The perimeter is $2L + 2B$, so

$$E(2L + 2B) = 2E(L) + 2E(B)$$
$$= 2 \times 26 + 2 \times 14 = 80,$$

$$\text{Var}(2L + 2B) = 2^2 \text{Var}(L) + 2^2 \text{Var}(B)$$
$$= 4 \times 4 + 4 \times 1 = 20.$$

(b)   The difference between length and width is $L - B$, so

$$\text{E}(L - B) = \text{E}(L) - \text{E}(B)$$
$$= 26 - 14 = 12,$$

$$\text{Var}(L - B) = 1^2 \text{Var}(L) + (-1)^2 \text{Var}(B)$$
$$= 1 \times 4 + 1 \times 1 = 5.$$

(c)*   The area is LB. Since L and B are independent,

$$\text{E}(LB) = \text{E}(L) \times \text{E}(B)$$
$$= 26 \times 14 = 364.$$

The result in part (b) for the variance is an example of a general rule which applies when variances are combined, namely that they are always added.

Example 3.2.4

If A and B are independent Poisson random variables with distributions $\text{Po}(a)$ and $\text{Po}(b)$, give reasons why the random variables   (a) $Y = 2A$,   (b) $Z = A - B$ are not Poisson random variables.

One property of Poisson random variables is that the mean is equal to the variance. The method is to show that Y and Z do not have this property, so they can't have a Poisson distribution.

(a)   Since $A \sim \text{Po}(a)$, $\text{E}(A) = a$ and $\text{Var}(A) = a$. So, from Section 3.1,

$$\text{E}(Y) = 2\text{E}(A) = 2a \quad \text{and} \quad \text{Var}(Y) = 2^2 \text{E}(A) = 4a.$$

Since $\text{E}(Y) \neq \text{Var}(Y)$, Y does not have a Poisson distribution.

(b)   From the results in this section,

$$\text{E}(Z) = 1\text{E}(A) + (-1)\text{E}(B) = a - b, \quad \text{and} \quad \text{Var}(Z) = 1^2 \text{E}(A) + (-1)^2 \text{E}(B) = a + b.$$

Since b can't be 0, $\text{E}(Z) \neq \text{Var}(Z)$, so Z does not have a Poisson distribution.

3.3 Linear relations involving more than one observation of a random variable

In Section 3.2 the separate random variables may represent different quantities, as in Examples 3.2.1 and 3.2.2, but they may also represent repeated observations of a single random variable. Suppose, for example, you make two observations of a random variable X where $\text{E}(X) = 3$ and $\text{Var}(X) = 4$. Denote these observations by X_1 and X_2. Then the expected values and variances of X_1 and X_2 will be equal to the corresponding values for X. It follows that

$$\text{E}(X_1 + X_2) = \text{E}(X_1) + \text{E}(X_2)$$
$$= \text{E}(X) + \text{E}(X)$$
$$= 3 + 3 = 6,$$

and, providing the observations are independent, that

$$\text{Var}(X_1 + X_2) = \text{Var}(X_1) + \text{Var}(X_2)$$
$$= \text{Var}(X) + \text{Var}(X)$$
$$= 4 + 4 = 8.$$

It is instructive to compare these results with the values for $E(2X)$ and $\text{Var}(2X)$ obtained using the results $E(aX) = aE(X)$ and $\text{Var}(aX) = a^2\text{Var}(X)$:

$$E(2X) = 2E(X) \qquad \text{and} \qquad \text{Var}(2X) = 2^2\text{Var}(X)$$
$$= 2 \times 3 = 6, \qquad\qquad\qquad = 2^2 \times 4 = 16.$$

The expected values for $X_1 + X_2$ and $2X$ are the same. This is not surprising since $2X = X + X$ and so $E(2X) = E(X) + E(X) = 6$. Why then do the variances of $X_1 + X_2$ and $2X$ differ? The answer is that $\text{Var}(2X)$ is not equal to $\text{Var}(X) + \text{Var}(X)$, because the variables X and X are not independent: they refer to the *same* observation of the same variable. This means that it is important to distinguish clearly between situations in which a single observation is multiplied by a constant and those in which several different observations of the same random variable are added.

Exercise 3

1   Random points with coordinates (x_i, y_i) lie on the line with equation $y = 4x - 3$. If $E(X) = 2$ and $\text{Var}(X) = 0.2$, find $E(Y)$ and $\text{Var}(Y)$.

2   A company pension scheme pays former employees a pension of $10 000 a year plus $500 for each year of service with the company. The directors want to predict the implications of the scheme for the year 2025, and use as a model for its surviving pensioners a distribution with a mean length of service of 17 years with standard deviation of 6 years. What does this predict for the mean and standard deviation of the pensions paid to former employees?

3   Intelligence test scores are scaled to produce a national mean of 100 with standard deviation 15. A new test is devised, and calibrated with a large-scale trial. This suggests that the new test produces national scores with mean 96 and standard deviation 12. What linear scaling formula should be used to convert raw marks R on the new test to scaled marks S? If a girl gets a raw mark of 120 on the new test, what should be her correct intelligence score?

4   The random variable X is the number of even numbers obtained when two ordinary fair dice are thrown. The random variable Y is the number of even numbers obtained when two fair pentagonal spinners, each numbered 1, 2, 3, 4, 5, are spun simultaneously.

Copy and complete the following probability distributions.

x	0	1	2
p	0.25		

y	0	1	2
p	0.36		

$x + y$	0	1	2	3	4
p	0.09				

Find $E(X)$, $\text{Var}(X)$, $E(Y)$, $\text{Var}(Y)$, $E(X + Y)$, $\text{Var}(X + Y)$ by using the probability distributions. Verify that $E(X + Y) = E(X) + E(Y)$ and $\text{Var}(X + Y) = \text{Var}(X) + \text{Var}(Y)$.

5 X and Y are independent random variables with probability distributions as shown.

x	1	2	3
p	0.4	0.2	0.4

y	0	1	2
p	0.3	0.5	0.2

You are given that $E(X) = 2$, $Var(X) = 0.8$, $E(Y) = 0.9$ and $Var(Y) = 0.49$.

The random variable T is defined as $2X - Y$. Find $E(T)$ and $Var(T)$ using the equations in Section 3.2.

Check your answers by completing the following probability distribution and calculating $E(T)$ and $Var(T)$ directly.

t	0	1	2	3	4	5	6
p	0.08						0.12

6 The independent random variables W, X and Y have means 10, 8 and 6 respectively and variances 4, 5 and 3 respectively.

Find $E(W + X + Y)$, $Var(W + X + Y)$, $E(2W - X - Y)$ and $Var(2W - X - Y)$.

7 A piece of laminated plywood consists of three pieces of wood of type A and two pieces of type B. The thickness of A has mean 2 mm and variance 0.04 mm^2. The thickness of B has mean 1 mm and variance 0.01 mm^2. Find the mean and variance of the thickness of the laminated plywood.

8 The random variable S is the score when an ordinary fair dice is thrown. The random variable T is the number of tails obtained when a fair coin is tossed once.

Find $E(S)$, $Var(S)$, $E(6T)$, $Var(6T)$, $E(S + 6T)$ and $Var(S + 6T)$.

9 The random variable X has the probability distribution shown in the table.

x	1	2	3	4
$P(X = x)$	$\frac{3}{8}$	$\frac{1}{4}$	$\frac{1}{4}$	$\frac{1}{8}$

Find the mean and variance of X.

Hence find the mean and variance of the distribution of the sum of three independent observations of X.

10 In American football, the quarterback's rating, Q, is calculated using the formula $Q = \frac{5}{6}(C + 5Y + 2.5 + 4T - 5I)$, where C, Y, T and I are variables which can be considered independent with means and variances as shown in the table.

Variable	C	Y	T	I
Mean	60.0	6.8	4.5	3.1
Variance	68.5	2.3	9.0	7.1

C, T and I are the percentage completions, touchdown passes and interceptions per pass attempt. Y is the number of yards gained divided by the passes attempted.

Find $E(Q)$ and $Var(Q)$.

11 The random variable Y which can only take the values 0 and 1 is called the Bernoulli distribution. Given that $P(1) = p$, show that $E(Y) = p$ and $\text{Var}(Y) = p(1 - p)$.

The binomial distribution can be considered to be a series of Bernoulli trials. That is,
$$X = Y_1 + Y_2 + \ldots + Y_n.$$
Show that $E(X) = np$ and $\text{Var}(X) = np(1 - p)$.

12 Let X_1, X_2, $\ldots$, X_n be n values randomly selected from a population, with mean μ and variance σ^2.

By considering $\overline{X} = \dfrac{X_1 + X_2 + \ldots + X_n}{n}$, show that $E(\overline{X}) = \mu$ and $\text{Var}(\overline{X}) = \dfrac{\sigma^2}{n}$.

4 Some properties of normal probability

Normal probability has some important properties which no other distributions have. When you have completed this chapter, you should

- know that any linear combination of normal random variables is also normal
- understand what is meant by the sampling distribution of the mean
- be able to apply this to problems involving the location of the mean of a normal population.

4.1 Linear combinations of random variables

When you have a random variable X with a particular kind of distribution, then aX and $X + b$ may or may not have the same kind of distribution.

For example, suppose that X is the uniform distribution of scores which you get when you roll a fair dice, with $P(X = x) = \frac{1}{6}$ for $X = 1, 2, 3, 4, 5, 6$. Then if you doubled the number of spots on each face of the dice, you would get a random variable $2X$ which also has a uniform distribution, with $P(X = x) = \frac{1}{6}$ for $X = 2, 4, 6, 8, 10, 12$. And if you put three extra dots on each face, you would get a random variable $X + 3$ with the uniform distribution $P(X = x) = \frac{1}{6}$ for $X = 4, 5, 6, 7, 8, 9$.

But if X has a Poisson distribution, then $2X$ and $X + 3$ do not have a Poisson distribution (see Example 3.2.4).

Similarly, when you have two independent random variables X and Y with a particular kind of distribution, then $X + Y$ may or may not have the same kind of distribution.

For example, it can be proved that, if X and Y are independent Poisson random variables, then $X + Y$ also has a Poisson distribution. But if X and Y have the uniform distribution of scores on a dice, then $X + Y$ has a triangular, not a uniform, distribution (see Higher Level Book 1 Section 33.1).

So it is a remarkable fact that, if X has a normal distribution, then so does $aX + b$; and if independent random variables X and Y are normally distributed, so is $X + Y$.

You can combine this with the rules given in Chapter 3,

$$E(aX + b) = aE(X) + b, \quad \text{Var}(aX + b) = a^2\text{Var}(X),$$

$$E(X + Y) = E(X) + E(Y), \quad \text{Var}(X + Y) = \text{Var}(X) + \text{Var}(Y) \quad \text{(for independent } X, Y)$$

to make precise statements about the distribution of $aX + b$ and $X + Y$.

> If $X \sim N(\mu, \sigma^2)$, then $aX + b$ is a normal random variable $N(a\mu + b, a^2\sigma^2)$.
>
> If X and Y are independent random variables $N(\mu_1, \sigma_1^2)$ and $N(\mu_2, \sigma_2^2)$, then $X + Y$ is a normal random variable $N(\mu_1 + \mu_2, \sigma_1^2 + \sigma_2^2)$.

If you like, you can combine these statements in one more general rule.

> If $X_1, X_2, \ldots, X_n$ are independent random variables, and if $X_i \sim N(\mu_i, \sigma_i^2)$,
> then $a_1 X_1 + a_2 X_2 + \ldots + a_n X_n + b$ has normal probability with distribution
> $N(a_1\mu_1 + a_2\mu_2 + \ldots + a_n\mu_n + b, a_1^2\sigma_1^2 + a_2^2\sigma_2^2 + \ldots + a_n^2\sigma_n^2)$.

Unfortunately the proofs are too difficult to give here.

Example 4.1.1

The random variable X is distributed $N(10, 5^2)$.

(a) What is the distribution of Y where $Y = 3X - 7$?

(b) Find the probability that a single observation of Y is less than 20.

(a) $E(Y) = E(3X - 7)$
$$= 3E(X) - 7$$
$$= 3 \times 10 - 7 = 23.$$

$\text{Var}(Y) = \text{Var}(3X - 7)$
$$= 3^2 \text{Var}(X)$$
$$= 9 \times 5^2 = 225.$$

Since Y is a linear function of X and X is normally distributed, Y is also normally distributed. Therefore $Y \sim N(23, 225)$.

(b)   You have various options for finding $P(Y < 20)$. With some calculators you can simply enter the boundaries of the interval, $-\infty$ and 20, and the values $\mu = 23$ and $\sigma = 15$, and find directly that $P(Y < 20) = 0.421$, correct to 3 significant figures.

You probably can't actually enter '$-\infty$' in your calculator, but any large negative number will do instead, since the areas under the tails of the normal probability density curve are negligible. To be completely safe, you can key in $-1\text{E}99$, which stands for -1×10^{99}.

Alternatively you can begin by standardising, writing

$$P(Y < 20) = P\left(Z < \frac{20 - 23}{\sqrt{225}}\right)$$
$$= P(Z < -0.2).$$

You can then either find this directly from the calculator program for the cumulative distribution function for $N(0,1)$, using boundaries $-\infty$ and -0.2, or you can use a table of areas under the standard normal curve. Since the table only gives values of $P(Z \leq z)$ for positive values of z, you have to use the symmetry properties of the $N(0,1)$ graph:

$$P(Z < -0.2) = P(Z > 0.2)$$
$$= 1 - P(Z \leq 0.2)$$
$$= 1 - 0.5793 \approx 0.421.$$

It is expected that most students will use a calculator to find normal probabilities, so the intermediate algebraic steps in calculations like this will usually be left out.

Example 4.1.2

The mass of an empty lift cage is 210 kg. If the masses (in kg) of adults are distributed as $N(70, 950)$, find the probability that the mass of the lift cage containing 10 adults chosen at random exceeds 1000 kg.

Let the mass in kg of an adult chosen at random be X. Then $X \sim N(70, 950)$.

Let the mass of the cage containing 10 adults be M. Then $M = 210 + X_1 + X_2 + \ldots + X_{10}$.

Assuming that the masses of the adults are independent, M will be normally distributed with mean and variance given by

$$
\begin{aligned}
E(M) &= E(210 + X_1 + X_2 + \ldots + X_{10}) \\
&= 210 + E(X_1) + E(X_2) + \ldots + E(X_{10}) \\
&= 210 + 10 \times E(X) \\
&= 210 + 10 \times 70 = 910,
\end{aligned}
$$

and
$$
\begin{aligned}
\text{Var}(M) &= \text{Var}(210 + X_1 + X_2 + \ldots + X_{10}) \\
&= \text{Var}(X_1) + \text{Var}(X_2) + \ldots + \text{Var}(X_{10}) \\
&= 10 \times \text{Var}(X) \\
&= 10 \times 950 = 9500.
\end{aligned}
$$

So $M \sim N(910, 9500)$.

To find $P(M > 1000)$, either use your calculator as to find the answer directly as described in Example 4.1.1, or standardise first to give

$$
\begin{aligned}
P(M > 1000) &= P\left(Z > \frac{1000 - 910}{\sqrt{9500}} \right) \\
&= P(Z > 0.9233\ldots) = 0.1779.
\end{aligned}
$$

The probability that the mass of the lift cage with 10 adults exceeds 1000 kg is 0.178, correct to 3 significant figures.

Example 4.1.3

An engineering company buys steel rods and steel tubes. Without heating the tubes so that they expand, an insufficient proportion of the rods will fit inside the tubes. Measured in centimetres, the internal diameter at room temperature of a randomly chosen tube is denoted by T, and the diameter at room temperature of a randomly chosen rod is denoted by R. It is given that $T \sim N(4.00, 0.10^2)$, that $R \sim N(4.02, 0.10^2)$, and that T and R are independent.

(a) Find the probability that a randomly chosen rod would fit inside a randomly chosen tube, without heating the tube.

(b) The tubes are heated so that the internal diameter of each tube increases by 5%. Find the probability that a randomly chosen rod fits inside a randomly chosen tube, after the tube has been heated.

(OCR)

(a) In order for a randomly chosen rod to fit inside a randomly chosen tube it is necessary that $T > R$. This inequality can also be expressed as $T - R > 0$, so the problem can be solved by considering the distribution of $T - R$. Since T and R are independent, $T - R$ is normally distributed with

$$E(T - R) = E(T) - E(R)$$
$$= 4.00 - 4.02 = -0.02,$$

and $\quad \text{Var}(T - R) = 1^2 \times \text{Var}(T) + (-1)^2 \times \text{Var}(R)$
$$= (0.10)^2 + (0.10)^2 = 0.02.$$

So $T - R \sim N(-0.02, 0.02)$.

A randomly chosen rod will fit inside a randomly chosen tube if $T - R > 0$.

$$P(T - R > 0) = P\left(Z > \frac{0 - (-0.02)}{\sqrt{0.02}} \right)$$
$$= P(Z > 0.1414\ldots)$$
$$= 0.4437\ldots\ .$$

The probability that a randomly chosen rod fits inside a randomly chosen tube at room temperature is 0.444, correct to 3 significant figures.

(b) When the tubes have been heated, their diameter is given by $1.05T$. A rod will now fit inside a tube provided that $1.05T - R > 0$.

$$E(1.05T - R) = 1.05E(T) - E(R)$$
$$= 1.05 \times 4.00 - 4.02$$
$$= 0.18.$$

$$\text{Var}(1.05T - R) = (1.05)^2 \text{Var}(T) + (-1)^2 \text{Var}(R)$$
$$= (1.05)^2 \times (0.1)^2 + (0.1)^2$$
$$= 0.021\,025.$$

Thus $1.05T - R \sim N(0.18, 0.021\,025)$.

$$P(1.05T - R > 0) = P\left(Z > \frac{0 - 0.18}{\sqrt{0.021\,025}} \right)$$
$$= P(Z > -1.241\ldots)$$
$$= 0.8927\ldots\ .$$

The probability that a randomly chosen rod fits inside a randomly chosen tube after the tube has been heated is 0.893, correct to 3 significant figures.

Exercise 4A

1   The heights of a population of male students are distributed normally with mean 178 cm and standard deviation 5 cm. The heights of a population of female students are distributed normally with mean 168 cm and standard deviation 4 cm. Find the probability that a randomly chosen female is taller than a randomly chosen male.

2   W is the mass of wine in a fully filled bottle, B is the mass of the bottle and C is the mass of the crate into which 12 filled bottles are placed for transportation, all in grams. It is given that $W \sim N(825, 15^2)$, $B \sim N(400, 10^2)$ and $C \sim N(1500, 20^2)$. Find the probability that a fully filled crate weighs less than 16.1 kg.

3   The times of four athletes for the 400 m are each distributed normally with mean 47 seconds and standard deviation 2 seconds. The four athletes are to compete in a 4×400 m relay race. Find the probability that their total time is less than 3 minutes.

4   The capacities of small bottles of perfume are distributed normally with mean 50 ml and standard deviation 3 ml. The capacities of large bottles of the same perfume are distributed normally with mean 80 ml and standard deviation 5 ml. Find the probability that the total capacity of three small bottles is greater than the total capacity of two large bottles.

5   The diameters of a consignment of bolts are distributed normally with mean 1.05 cm and standard deviation 0.1 cm. The diameters of a consignment of nuts are distributed normally with mean 1.1 cm and standard deviation 0.1 cm. Find the probability that a randomly chosen bolt will not fit inside a randomly chosen nut.

6   The amount of black coffee dispensed by a drinks machine is normally distributed with mean 200 ml and standard deviation 5 ml. If a customer requires white coffee, milk is also dispensed. The amount of milk is distributed normally with mean 20 ml and standard deviation 2 ml. Find the probability that the total amount of liquid dispensed when a customer chooses white coffee is less than 210 ml.

7   Given that $X \sim N(\mu, 10)$, $Y \sim N(12, \sigma^2)$ and $3X - 4Y \sim N(0, 234)$, find μ and σ^2.

4.2   Sampling from a normal population

A researcher for a Ruritanian sock manufacturer states that the average man's foot has a length of 28 cm.

How does she know?

She certainly didn't measure the foot of every Ruritanian man and work out the mean. On the other hand, she didn't just measure the foot of the first Ruritanian man she met in the street, and assume that it was typical of every man in the country.

A practical procedure which comes between these extremes is to select, say, 100 men and take the mean of their foot lengths. They must, of course, be selected randomly; you wouldn't choose too many basketball players or policemen because their feet are likely to be larger than average. There are special techniques to make sure that the selection is free of bias.

But why is it better to take the mean of 100 foot lengths than just one? How many should she select? Would 10 be enough, or should she take 1000? These are questions which can be answered by the results in Section 4.1.

At this stage it is useful to introduce some technical language. The aim of the researcher is to throw light on the complete set of foot lengths of all the men in Ruritania. This set of lengths is called the **population**. (Notice that, in statistics, the population is the set of measurements, *not* the men themselves.) From this population she randomly selects a **sample**. The mean of the sample is then used to give information about the mean of the population.

It is a reasonable assumption that, for the whole population, the lengths can be modelled by a random variable X with a normal distribution $N(\mu,\sigma^2)$. The values of μ and σ^2 are unknown; in fact the researcher is using the sample to help her decide the value of μ.

When the researcher selects her sample, she obtains 100 different lengths $X_1, X_2, \ldots, X_{100}$ from the population. Each of these is an individual random variable whose distribution is the same as that of X; and the random method of selection should ensure that these are independent. (For example, she would not deliberately select two brothers. But there is always the remote possibility that two brothers could be selected by chance, so brothers would not be deliberately excluded either.) She will then take the mean of the 100 lengths, which produces another random variable $\overline{X}$, defined as

$$\overline{X} = \frac{X_1 + X_2 + \ldots + X_{100}}{100}$$
$$= \tfrac{1}{100}X_1 + \tfrac{1}{100}X_2 + \ldots + \tfrac{1}{100}X_{100}.$$

So, from the result in the shaded box in Section 4.1, $\overline{X}$ has normal probability

$$N\!\left(\tfrac{1}{100}\mu + \tfrac{1}{100}\mu + \ldots + \tfrac{1}{100}\mu, \tfrac{1}{10\,000}\sigma^2 + \tfrac{1}{10\,000}\sigma^2 + \ldots + \tfrac{1}{10\,000}\sigma^2\right),$$

which is $N\!\left(\mu, \tfrac{1}{100}\sigma^2\right)$. Therefore $\overline{X}$, the mean of the sample, has the same mean as that of the population. But its standard deviation is $\tfrac{1}{10}\sigma$, which is $\tfrac{1}{10}$ of the standard deviation of the population.

This is called the **sampling distribution of the mean**. It is illustrated in Fig. 4.1. The probability density graph on the left is that of $N(\mu,\sigma^2)$, the distribution of the population. The graph on the right is that of $N\!\left(\mu, \tfrac{1}{100}\sigma^2\right)$, the distribution of the sample mean.

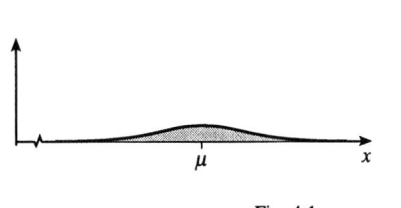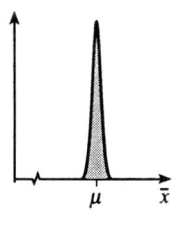

Fig. 4.1

For the theory to be completely valid, the sampling should be carried out 'with replacement'. That is, the method of selection should allow the possibility that the same man's measurement is used more than once. But amongst several million Ruritanian men, the probability of this happening is very small.

The graphs show the advantage of using a sample rather than an individual measurement to make an estimate of μ. Consider the 'two standard deviations' regions which contain more than 95% of the probability of the random variables X and $\overline{X}$. By taking a sample the width of this region has been reduced to one-tenth of its original value.

In general, if a random sample of size n is drawn from a population $N(\mu, \sigma^2)$, then by a similar argument the sampling distribution of the mean is

$$N\left(\frac{1}{n}\mu + \frac{1}{n}\mu + \dots + \frac{1}{n}\mu, \frac{1}{n^2}\sigma^2 + \frac{1}{n^2}\sigma^2 + \dots + \frac{1}{n^2}\sigma^2\right), \quad \text{which is} \quad N\left(\mu, \frac{\sigma^2}{n}\right).$$

So if the researcher were to take a sample of 400 foot lengths rather than 100, she would only halve the width of the sampling distribution. Her estimate of μ would certainly be more reliable, but perhaps not enough more to justify the extra expense.

> If the distributions of $X_1, X_2, \dots, X_n$ are independent $N(\mu, \sigma^2)$ variables, then
>
> the distribution of $\overline{X}$ is $N\left(\mu, \frac{\sigma^2}{n}\right)$.

Example 4.2.1

The mass of a randomly chosen male student in Year 10 at a large secondary school may be modelled by a normal distribution with mean 55 kg and standard deviation 2.2 kg. Four students are chosen at random from this year group. Calculate the probability that the mean mass of the four students is
(a) less than 58 kg,   (b) between 52 kg and 57.5 kg.

Let M_1, M_2, M_3 and M_4 be the masses of four randomly chosen male Year 10 students. Then

$$M_i \sim N(55, 2.2^2) \text{ for } i = 1, 2, 3, 4.$$

Therefore $\overline{M} \sim N\left(55, \frac{2.2^2}{4}\right) = N(55, 1.21)$.

Note that, as the distribution for each M_i is normal, the distribution of $\overline{M}$ is normal.

If you are using normal probability tables, standardise by letting $Z = \dfrac{\overline{M} - 55}{1.1}$, then $Z \sim N(0,1)$.

(a) $P(\overline{M} < 58) = P\left(Z < \dfrac{58 - 55}{1.1}\right)$

  $= P(Z < 2.727...)$

  $= 0.997$, correct to 3 significant figures.

(b) $P(52 < \overline{M} < 57.5) = P\left(\dfrac{52 - 55}{1.1} < Z < \dfrac{57.5 - 55}{1.1}\right)$

  $= P(-2.727... < Z < 2.272...)$

  $= 0.985$, correct to 3 significant figures.

If you are using a calculator that gives the answer directly, simply key in $\mu = 55$ and $\sigma = \sqrt{1.21}$, with boundaries $-\infty$ and 58 for part (a), and 52 and 57.5 for part (b).

Example 4.2.2

A second sample of size n is chosen from the male Year 10 students. How large does n have to be for there to be at most a 2% chance that the mean mass of the sample differs from the mean mass of the population by more than $0.6\,\text{kg}$?

Let $M_1, M_2, \ldots, M_n$ be the masses of n randomly chosen Year 10 male students. Then

$$\overline{M} \sim \text{N}\!\left(55, \frac{2.2^2}{n}\right).$$

Then the minimum value of n is required for which

$$P(54.4 \le \overline{M} \le 55.6) \ge 0.98.$$

Standardising this gives

$$P\!\left(\frac{-0.6}{2.2/\sqrt{n}} \le Z \le \frac{0.6}{2.2/\sqrt{n}}\right) \ge 0.98.$$

This is illustrated in Fig. 4.2. The shaded area has to be greater than 0.98, so each of the unshaded areas under the graph is less than 0.01. You therefore want to find n such that

$$P\!\left(Z \le \frac{0.6\sqrt{n}}{2.2}\right) \ge 0.99.$$

Using a calculator or a table of inverse normal probabilities, the number z such that $P(Z \le z) = 0.99$ is $2.326\ldots$. So n must satisfy the inequality

$$\frac{0.6\sqrt{n}}{2.2} \ge 2.326\ldots,$$

$$n \ge \left(\frac{2.326\ldots \times 2.2}{0.6}\right)^2 = 72.7\ldots.$$

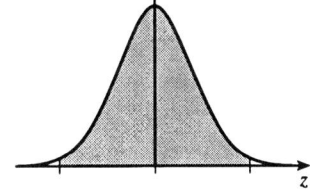

Fig. 4.2

Since n must be an integer, it has to be greater than or equal to 73. So the sample needs to contain at least 73 students for its mean mass to approximate the population's mean mass with the accuracy and certainty desired.

Exercise 4B

1   The body length of a species of ant is normally distributed with mean 3.1 mm and standard deviation 0.2 mm.

(a) What is the probability that an ant of this species chosen at random has a body length greater than 3.12 mm?

(b) What is the probability that the mean body length of a sample of
(i) 9 ants is greater than 3.12 mm,   (ii) 100 ants is greater than 3.12 mm?

2 Random samples of three are drawn from a population of beetles whose lengths have a normal distribution with mean 2.4 cm and standard deviation 0.36 cm. The mean length $\overline{X}$ is calculated for each sample.

(a) State the distribution of $\overline{X}$, giving the values of its parameters.

(b) Find $P(\overline{X} > 2.5)$.

3 The masses of kilogram bags of flour produced in a factory have a normal distribution with mean 1.005 kg and standard deviation 0.0082 kg. A shelf in a store is loaded with 22 of these bags, assumed to be a random sample.

(a) Find the probability that a randomly chosen bag has mass less than 1 kg.

(b) Find the probability that the mean mass of the 22 bags is less than 1 kg.

4 An insurance company offers a pension scheme for retiring executives of large firms. Each firm pays the insurance company $1 200 000 on the day that its executive retires. The insurance company will pay the executives a pension of $10 000 a month for as long as they live. Past experience suggests that the length of retirement enjoyed by executives is normally distributed with a mean of 9 years and a standard deviation of 3 years. If in one year the insurance company sets up 400 such contracts, what is the probability that it will make a loss? (Neglect interest on capital, staff costs, etc.)

One firm decides not to join the scheme, but to pay its retiring executive a pension directly from its own resources. What is the probability that the firm will be better off than if it had joined the scheme?

5 A drink is sold in bottles of nominal capacity 250 ml. A shop takes delivery of 100 bottles. The bottles are filled by a machine which dispenses the drink in quantities which are normally distributed with mean 251.5 ml and standard deviation 5 ml. Find the probabilities that

(a) the delivery does not contain any bottles which are underfilled,

(b) the mean content of all the bottles in the batch is not below the stated capacity.

6 The mean of a random sample of 500 observations of the random variable X, where $X \sim N(25,18)$ is denoted by $\overline{X}$. Find the value of a for which $P(\overline{X} < a) = 0.25$.

7 The life of Powerlong batteries, sold in packs of 6, may be assumed to have a normal distribution with mean 32 hours and standard deviation σ hours. Find the value of σ so that for one box in 100 (on average) the mean life of the batteries is less than 30 hours.

5 Hypothesis testing

An important aim of statistics is to make informed decisions based on data. This chapter uses probability to test hypotheses. When you have completed it, you should

- understand the nature of a hypothesis test
- be able to formulate a null hypothesis and an alternative hypothesis
- understand the difference between a one-tail and a two-tail test
- understand the terms 'critical values', 'significance level', 'rejection region', 'acceptance region' and 'test statistic'
- be able to carry out a hypothesis test of a population mean for a sample drawn from a normal distribution of known variance.

5.1 An introductory example

Over a number of years a primary school has recorded the reading ages of children at the beginning and end of each academic year. The teachers have found that during the third year the increase in reading age is normally distributed with mean 1.14 years and standard deviation 0.16 years. This year they are going to trial a new reading scheme: other schools have tried this scheme and found that it led to a greater increase in reading age. At the end of the year the teachers will use the mean increase in reading age, $\bar{x}$, of the 40 third-year children to help them answer the question: 'Does the new reading scheme give better results than the old one in our school?'

The difficulty in answering this question lies in the fact that each child progresses at a different rate so that different values of $\bar{x}$ will be obtained for different groups of children. It is easy to check whether $\bar{x}$ is greater than 1.14 years. It is not easy to know whether a value of $\bar{x}$ greater than 1.14 years reflects the effectiveness of the new scheme or is just due to random variation between children.

This chapter explores how statistics can be used to arrive at such a decision. The following sections break down the process into several stages.

5.2 Null and alternative hypotheses

There are two theories about how the new reading scheme performs in this particular school. The first is that using the new scheme makes no difference. This theory is called a **null hypothesis**. It is denoted by the symbol H_0. In this example, where the mean of the past increases in reading age is 1.14 years, the null hypothesis can be expressed by $H_0 : \mu = 1.14$. Note that H_0 proposes a single value for the population mean, μ, which is based on past experience.

A 'hypothesis' is a theory which is assumed to be true unless evidence is obtained which indicates otherwise. 'Null' means 'nothing' and the term 'null hypothesis' means a 'theory of no change', that is 'no change' from what would be expected from past experience.

The other theory is that the new reading scheme is more effective than the old one, that is that the population mean will increase. This is called the **alternative hypothesis** and is given the symbol H_1. So the alternative hypothesis in this case is $H_1 : \mu > 1.14$. The alternative hypothesis proposes the way in which μ will have changed if the new reading scheme is more effective than the old one.

The procedure which is used to decide between these two opposing theories is called a **hypothesis test** or sometimes a **significance test**. In this example the test will be **one-tail** because the alternative hypothesis proposes a change in the mean in only one direction, in this case an increase. It is also possible to have a one-tail test in which the alternative hypothesis proposes a decrease in the mean. Tests in which the alternative hypothesis suggests a difference in the mean in either direction are called **two-tail** tests.

Example 5.2.1

For the following situations give null and alternative hypotheses and say whether a hypothesis test would be one-tail or two-tail.

(a) In the past an athlete has run 100 metres in 10.3 seconds on average. He has been following a new training programme which he hopes will decrease the time he takes to run 100 metres. He is going to time himself on his next six runs.

(b) The bags of sugar coming off a production line have masses which vary slightly but which should have a mean value of 1.01 kg. A sample is to be taken to test whether there has been any change in the mean.

(c) The mean volume of liquid in bottles of lemonade should be at least 2 litres. A sample of bottles is taken in order to test whether the mean volume has fallen below 2 litres.

(a) The null hypothesis proposes a single value for μ, $H_0: \mu = 10.3$, based on the athlete's past performance. The alternative hypothesis proposes how μ might have decreased, $H_1: \mu < 10.3$. This is a one-tail test.

(b) The null hypothesis proposes a single value for μ, $H_0: \mu = 1.01$, based on the mass which the bags should have. The alternative hypothesis proposes how μ might have changed, $H_1: \mu \neq 1.01$. This is a two-tail test.

(c) The null hypothesis proposes that μ should be at least 2, that is $\mu \geq 2$. However, you will see later in the chapter that a single value of μ is needed in order to carry out the calculation in a hypothesis test. So, in this example, the null hypothesis $H_0: \mu = 2$ is taken. This null hypothesis satisfies the criterion that μ is at least 2. A sample is taken in order to test whether the mean has fallen below 2 so the alternative hypothesis is $H_1: \mu < 2$. This is a one-tail test.

For a hypothesis test on the population mean, μ, the null hypothesis, H_0, proposes a value, μ_0, for μ,

$$H_0: \mu = \mu_0.$$

The alternative hypothesis, H_1, suggests the way in which μ might differ from μ_0. H_1 can take three forms:

$H_1: \mu < \mu_0$, a one-tail test for a decrease

$H_1: \mu > \mu_0$, a one-tail test for an increase

$H_1: \mu \neq \mu_0$, a two-tail test for a difference.

In the following situations, state suitable null and alternative hypotheses involving a population with mean μ. You will need some of your answers in Exercise 5B.

1　Bars of Choco are claimed by the manufacturer to have a mean mass of 102.5 grams. A test is carried out to see whether the mean mass of Choco bars is less than 102.5 grams.

2　The mean factory assembly time for a particular electronic component is 84 seconds. It is required to test whether the introduction of a new procedure results in a different assembly time.

3　In a report it was stated that the average age of all hospital patients was 53 years. A newspaper believes that this figure is an underestimate.

4　The manufacturer of a certain battery claims that it has a mean life of 30 hours. A suspicious customer wishes to test the claim.

5　A large batch of capacitors is judged to be satisfactory by an electronics factory if the mean capacitance is at least 5 microfarads. A test is carried out on a batch to determine whether it is satisfactory.

5.3　Critical values

Once you have decided on your null and alternative hypotheses the next step is to devise a rule for choosing between them. Look again at the reading scheme example. The rule will be based on the sample mean, $\bar{x}$. The teachers are only interested in the new scheme if it improves the average increase in reading age and so only values of $\bar{x}$ greater than 1.14 might lead them to drop the old scheme in favour of the new one. Initially, you might think that *any* value of $\bar{x}$ which is greater than 1.14 years would show that the new scheme is more effective. A little more thought shows that is too simple a rule.

It is possible to obtain a sample mean $\bar{x}$ that is greater than 1.14 even if the new reading scheme is not effective at all. To see this suppose that there is no difference between the new reading scheme and the old one. Then both μ and σ will be the same under the two schemes and X will be distributed as $N(1.14, 0.16^2)$.

You have already seen in Section 4.2 that if $X \sim N(\mu, \sigma^2)$, then $\overline{X} \sim N\left(\mu, \dfrac{\sigma^2}{n}\right)$; so for samples consisting of 40 children, the mean, $\overline{X}$, will be distributed as $N\left(1.14, \dfrac{0.16^2}{40}\right)$.

Fig. 5.1 shows the distribution of $\overline{X}$.

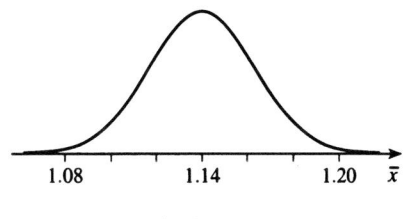

Fig. 5.1

You can see from this diagram that there is a probability of $\frac{1}{2}$ that the sample mean will be greater than 1.14 even though you assumed that there is no change in the population mean.

How big does the sample mean have to be before you can conclude that the population mean is likely to have increased from 1.14? Most people would agree that if a sample mean of 2.00 is obtained then it is unlikely that the population mean is still 1.14; but what about a sample mean of 1.19? One way of tackling this problem is to divide the possible outcomes into two regions: the **rejection (or critical) region** and the **acceptance region.**

The rejection region will contain values at the top end of the distribution in Fig. 5.1.

If the sample mean is in the rejection region, you reject H_0 in favour of H_1: you conclude that the population mean has increased. If the sample mean is in the acceptance region, you do not reject H_0: there is insufficient evidence to say that the new reading scheme is more effective.

The rejection region is chosen so that it is 'unlikely' for the sample mean to fall in the rejection region when H_0 is true. It is a matter of opinion what you mean by 'unlikely' but a common convention among statisticians is that an event which has a probability of 0.05, that is 1 in 20, or less is 'unlikely'.

Fig. 5.2 shows the rejection region and the acceptance region for the children's reading scheme example. The value, c, which separates the rejection and acceptance regions is called a **critical value**. It can be calculated as follows.

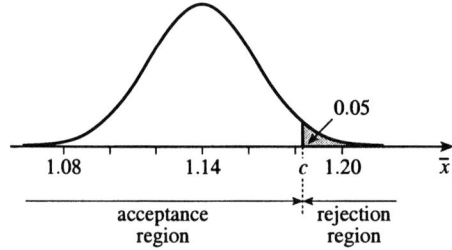

Fig. 5.2

Begin by finding the value of the standardised normal variable, z, such that $P(Z \le z) = 0.95$. Using a calculator or tables, this is $z = 1.645$, correct to 3 decimal places.

The values of the standardised and original variables are related by $z = \dfrac{\bar{x} - \mu}{\sqrt{\dfrac{\sigma^2}{n}}}$ or $z = \dfrac{\bar{x} - \mu}{\dfrac{\sigma}{\sqrt{n}}}$, where $Z \sim N(0,1)$.

Substituting in this equation gives $1.645 = \dfrac{c - 1.14}{\dfrac{0.16}{\sqrt{40}}}$.

Rearranging gives $c = 1.645 \times \dfrac{0.16}{\sqrt{40}} + 1.14 = 1.18$, correct to 3 significant figures.

The rejection region is given by $\bar{X} \ge 1.18$ years.

At the end of the year the observed value for the sample mean was $\bar{x} = 1.19$ years. Since this value is in the rejection region, you can conclude that the observed result is unlikely to be explained by random variation; it is more likely to be due to an increase in the population mean. This suggests that the new reading scheme does give better results than the old one.

In this example a decision is made by considering the value of the sample mean. The sample mean, $\bar{X}$, is called the **test statistic** for this hypothesis test. The rejection region was defined so that the probability of the test statistic falling in it, *if* H_0 *is true*, is at most 0.05, or 5%. This probability is called the **significance level** of the test. It gives the probability of rejecting H_0 when it is in fact true. In this example it gives the probability of concluding that the new reading scheme is better even when it is not. You might feel that this is too high a risk of being wrong and choose instead to use a significance level of, say, 0.01, or 1%.

Example 5.3.1

Find the rejection region for a test at the 1% significance level for the children's reading scheme example.

Now $P(Z \le z) = 0.99$ gives $z = 2.326$. Substituting into $z = \dfrac{\bar{x} - \mu}{\dfrac{\sigma}{\sqrt{n}}}$ gives

$$2.326 = \frac{c - 1.14}{\dfrac{0.16}{\sqrt{40}}}.$$

Rearranging gives $c = 1.20$, correct to 3 significant figures.

The rejection region is $\bar{X} \ge 1.20$.

The observed value of 1.19 years is no longer in the rejection region and so H_0 is not rejected at the 1% significance level.

You may feel that it is unsatisfactory that the result of a hypothesis test should depend on the significance level chosen. This point is discussed in more detail in Section 9.2.

In a two-tail test the rejection region has two parts, because both high and low values of $\bar{X}$ are unlikely if the null hypothesis is true. Example 5.3.2 illustrates this situation.

Example 5.3.2

In the past a machine has produced rope which has a breaking load which is normally distributed with mean 1000 newtons and standard deviation 21 newtons. A new process has been introduced. To test whether the mean breaking load has changed a sample of 50 pieces of rope is taken, the breaking strain of each piece measured and the mean calculated.

(a) Define suitable null and alternative hypotheses for testing whether the breaking load has changed.

(b) Taking the sample mean as the test statistic, find the rejection region for $\bar{X}$ for a hypothesis test at the 5% significance level.

(c) The sample mean for the 50 pieces of rope was 1003 newtons. What can you deduce?

(a) The null hypothesis states the value which the mean breaking load should take, $H_0 : \mu = 1000$.

The alternative hypothesis states how μ might have changed, $H_1 : \mu \ne 1000$.

(b) If H_0 is true, then the sample mean $\bar{X} \sim N\left(1000, \dfrac{21^2}{50}\right)$. Fig. 5.3 shows the distribution of $\bar{X}$ with the rejection and acceptance regions. This is a two-tail test so there are two critical values labelled c_1 and c_2.

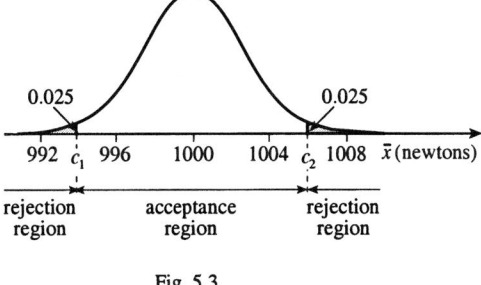

Fig. 5.3

To find the upper critical value, use $P(Z \le z) = 0.975$, since the 0.05 probability is split equally between the two 'tails' of the distribution. The required value of z is 1.960, correct to 3 decimal places.

Substituting into $z = \dfrac{\bar{x} - \mu}{\dfrac{\sigma}{\sqrt{n}}}$ gives $1.960 = \dfrac{c_2 - 1000}{\dfrac{21}{\sqrt{50}}}$.

Rearranging gives $c_2 = 1006$, to the nearest integer. By symmetry, $c_1 = 994$.

So the rejection region is $\overline{X} \leq 994$ and $\overline{X} \geq 1006$.

(c)   The observed sample mean of $\bar{x} = 1003$ is not in the rejection region. There is not enough evidence to say that the mean has changed and it can be concluded that the new process is satisfactory.

Note that the conclusion to a hypothesis test should always be given in context.

Here is a summary of the terms introduced in this section, followed by a list of the steps involved in carrying out a hypothesis test.

The **test statistic** is calculated from the sample. Its value is used to decide whether the null hypothesis, H_0, should be rejected.

The **rejection** (or **critical**) **region** gives the values of the test statistic for which the null hypothesis, H_0, is rejected.

The **acceptance region** gives the values of the test statistic for which the null hypothesis, H_0, is not rejected.

The boundary value(s) of the rejection region is (are) called the **critical value(s)**.

The **significance level** of a test gives the probability of the test statistic falling in the rejection region when H_0 is true.

If H_0 is rejected, then H_1 is automatically accepted.

To carry out a hypothesis test:

Step 1   Define the null and alternative hypotheses.

Step 2   Decide on a significance level.

Step 3   Determine the critical value(s).

Step 4   Calculate the test statistic.

Step 5   Decide on the outcome of the test depending on whether the value of the test statistic is in the rejection or the acceptance region.

Step 6   State the conclusion in words.

Exercise 5B

In the following questions, the rejection (critical) regions should be found in terms of the sample mean, $\overline{X}$.

1   The random variable X has a normal distribution, $N(\mu, 4)$. A test of the null hypothesis $\mu = 10$ against the alternative hypothesis $\mu > 10$ is carried out, at the 5% significance level, using a random sample of 9 observations of X. The rejection region is found to be $\overline{X} \geq 11.10$.

State the conclusion of the test in the following cases.

(a) $\overline{X} = 12.3$ (b) $\overline{X} = 8.6$

2   The random variable Y has a normal distribution, $N(\mu, 9)$. A test of the null hypothesis $\mu = 15$ against the alternative hypothesis $\mu < 15$ is carried out at the 10% significance level, using a random sample of 16 observations. Show that the rejection region is $\overline{Y} < 14.04$.

State the conclusion of the test in the following cases.

(a) $\overline{Y} = 15.5$ (b) $\overline{Y} = 12.7$

3   The random variable X has a normal distribution, $N(\mu, 25)$. A test of the null hypothesis $\mu = 20$ against the alternative hypothesis $\mu \neq 20$ is carried out at the 5% significance level, using a random sample of 4 observations. Show that the rejection region is $\overline{X} < 15.1$ and $\overline{X} > 24.9$.

State the conclusion of the test in the following cases.

(a) $\overline{X} = 17$ (b) $\overline{X} = 13$ (c) $\overline{X} = 30$

4   For the situation in Exercise 5A Question 1, a random sample of 12 bars had a mean mass of 101.4 g. Test, at the 5% significance level, whether the mean mass of all Choco bars is less than 102.5 g, assuming that the mass of a Choco bar is normally distributed with standard deviation 1.7 g.

5   For the situation in Exercise 5A Question 2, a random sample of 40 components had mean assembly time 81.2 s. Assuming that the assembly time of a component has a normal distribution with standard deviation 6.1 s, carry out a test at the 5% significance level of whether the mean for all components differs from 84.0 s.

6   Referring to Exercise 5A Question 5, a random sample of 6 capacitors was selected from the batch. Their capacitances were measured in microfarads with the following results.

      5.12   4.81   4.79   4.85   5.04   4.61

Assuming that the capacitances have a normal distribution with standard deviation 0.35 microfarads, test, at the 2% significance level, whether the batch is satisfactory.

7   The blood pressure of a group of hospital patients with a certain type of heart disease has mean 85.6. A random sample of 25 of these patients volunteered to be treated with a new drug and a week later their mean blood pressure was found to be 70.4. Assuming a normal distribution with standard deviation 15.5 for blood pressures, and using a 1% significance level, test whether the mean blood pressure for all patients treated with the new drug is less than 85.6.

8   Two-litre bottles of a brand of spring water are advertised as containing 6.8 mg of magnesium. In a random sample of 10 of these bottles the mean amount of magnesium was found to be 6.92 mg. Assuming that the amounts of magnesium are normally distributed with standard deviation 0.18 mg, test whether the mean amount of magnesium in all similar bottles differs significantly from 6.8 mg. Use a 5% significance level.

9   The lives of a certain make of battery have a normal distribution with mean 30 h and variance 2.54 h^2. When a large consignment of these batteries is delivered to a store the quality control manager tests the lives of 8 randomly chosen batteries. The mean life was 28.8 h. Test whether there is cause for complaint. Use a 3% significance level.

10   The birth weights of babies born in a certain large hospital maternity unit during the year 2006 had a normal distribution with mean 3.21 kg and standard deviation 0.73 kg. During the first week of August, there were 24 babies born with a mean weight of 3.17 kg. Using a 5% significance level, test whether the sample is likely to differ from a sample chosen at random from the year's births at the hospital.

5.4   Standardising the test statistic

In Exercise 5B the rejection region for each question was different and you had to find it before you could obtain the result of the hypothesis test. You may have spotted that the calculation could be shortened by standardising the value of $\overline{X}$ using

$$Z = \frac{\overline{X} - \mu}{\sqrt{\dfrac{\sigma^2}{n}}} = \frac{\overline{X} - \mu}{\dfrac{\sigma}{\sqrt{n}}}$$

and taking Z as the test statistic. For a given type of test, one-tail or two-tail, at a given significance level the rejection region for Z will always be the same.

For example, Fig. 5.4 illustrates the rejection region of Z for a two-tail test at the 5% significance level. The upper critical value is obtained from $P(Z \le z_1) = 0.975$, giving $z_1 = 1.960$, correct to 3 decimal places and, by symmetry, the lower critical value, $z_2 = -1.960$. Thus the rejection region is $Z \ge 1.960$ and $Z \le -1.960$, which you can write more compactly as $|Z| \ge 1.960$.

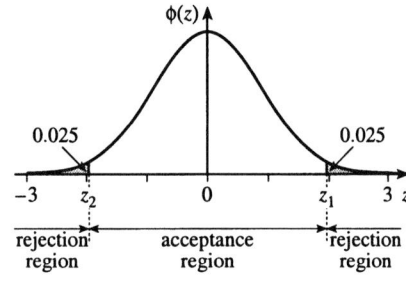

Fig. 5.4

The following examples illustrate this approach.

Example 5.4.1

A test of mental ability has been constructed so that, for adults in the UK, the test score is normally distributed with mean 100 and standard deviation 15. A doctor wishes to test whether sufferers from a particular disease differ in mean from the general population in their performance on this test. She chooses a random sample of 10 sufferers. Their scores on the test are

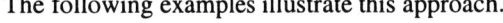

        119   131   95   107   125   90   123   89   103   103.

Carry out a test at the 5% significance level to test whether sufferers from the disease differ from the general population in the way in which they perform at this test.   (OCR, adapted)

The null and alternative hypotheses are $H_0: \mu = 100$ and $H_1: \mu \ne 100$ respectively, where μ is the mean score in a test of mental ability.

This is a two-tail test at the 5% significance level. As explained above, the rejection region for the test statistic, Z, is $|Z| \geq 1.960$.

Under H_0, $\overline{X} \sim N\left(100, \frac{15^2}{10}\right)$.

'Under H_0' is another way of saying 'If H_0 is true'.

For this sample,

$$\bar{x} = \tfrac{1}{10}(119 + 131 + 95 + 107 + 125 + 90 + 123 + 89 + 103 + 103) = 108.5.$$

When $\bar{x} = 108.5$, $z = \dfrac{\bar{x} - \mu}{\frac{\sigma}{\sqrt{n}}} = \dfrac{(108.5 - 100)}{\frac{15}{\sqrt{10}}} = 1.792$, correct to 3 decimal places.

The observed value of Z, 1.792, is not in the rejection region, $|Z| \geq 1.960$, so H_0 is not rejected. There is insufficient evidence, at the 5% significance level, to suggest that sufferers from this disease differ from the general population in their performance on the test.

Example 5.4.2

A manufacturer claims that its light bulbs have a lifetime which is normally distributed with mean 1500 hours and standard deviation 30 hours. A shopkeeper suspects that the bulbs do not last as long as is claimed because he has had a number of complaints from customers. He tests a random sample of six bulbs and finds that their lifetimes are 1472, 1486, 1401, 1350, 1511, 1591 hours. Is there evidence, at the 1% significance level, that the bulbs last a shorter time than the manufacturer claims?

The null and alternative hypotheses are $H_0: \mu = 1500$ and $H_1: \mu < 1500$ respectively, where μ is the mean lifetime of a light bulb in hours.

This is a one-tail test for a decrease at the 1% level. Fig. 5.5 shows the rejection region for Z. The critical value is obtained from $P(Z \leq z) = 0.01$, giving $z = -2.326$ and the rejection region is $Z \leq -2.326$.

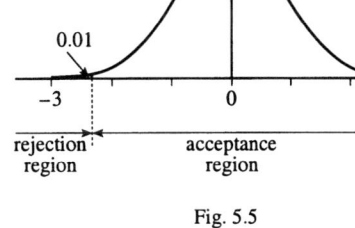

Fig. 5.5

Under H_0, $\overline{X} \sim N\left(1500, \frac{30^2}{6}\right)$.

For this sample,

$$\bar{x} = \tfrac{1}{6}(1472 + 1486 + 1401 + 1350 + 1511 + 1591) = 1468.5.$$

When $\bar{x} = 1468.5$, $z = \dfrac{\bar{x} - \mu}{\frac{\sigma}{\sqrt{n}}} = \dfrac{1468.5 - 1500}{\frac{30}{\sqrt{6}}} = -2.572$, correct to 3 decimal places.

The observed value of Z, −2.572, is in the rejection region, $Z \leq -2.326$. There is evidence, at the 1% significance level, that the manufacturer's bulbs do not last as long as claimed.

You can generalise this method as follows:

The test statistic Z can be used to test a hypothesis about a population mean, $H_0: \mu = \mu_0$, for samples drawn from a normal distribution of known variance σ^2. For a sample of size n, the value of Z is given by

$$z = \frac{\bar{x} - \mu}{\sqrt{\frac{\sigma^2}{n}}} = \frac{\bar{x} - \mu}{\frac{\sigma}{\sqrt{n}}}.$$

The rejection region for Z depends on H_1 and the significance level used. The critical values, correct to 3 decimal places, for some commonly used rejection regions are given below.

Significance level	Two-tail $H_1: \mu \neq \mu_0$	One-tail $H_1: \mu > \mu_0$	One-tail $H_1: \mu < \mu_0$
10%	± 1.645	1.282	-1.282
5%	± 1.960	1.645	-1.645
2%	± 2.326	2.054	-2.054
1%	± 2.576	2.326	-2.326

Exercise 5C

1   Cans of lemonade are filled by a machine which is set to dispense an amount which is normally distributed with mean 330 ml and standard deviation 2.4 ml. A quality control manager suspects that the machine is over-dispensing and tests a random sample of 8 cans. The volumes of the contents, in ml, are as follows:

329   327   331   326   334   343   328   339.

Test, at the $2\frac{1}{2}\%$ significance level, whether the manager's suspicion is justified.

2   The masses of loaves from a certain bakery have a normal distribution with mean μ grams and standard deviation σ grams. When the baking procedure is under control, $\mu = 508$ and $\sigma = 18$. A random sample of 25 loaves from a day's output had a total mass of 12 554 grams. Does this provide evidence at the 10% significance level that the process is not under control?

3   A machine produces elastic bands with breaking tension T newtons, where $T \sim N(45.1, 19.0)$. On a certain day, a random sample of 50 bands was tested and found to have a mean breaking tension of 43.4 newtons. Test, at the 4% significance level, whether this indicates a change in the mean breaking tension.

4   The cholesterol level of healthy males under the age of 21 is normally distributed with mean 160 and standard deviation 10. A random sample of 200 university students, all under age 21, had a mean cholesterol level of 161.8. Test, at the 1% significance level, whether all male university students under age 21 have a mean cholesterol level greater than 160.

5   The mean and standard deviation of the number of copies of *The Daily Courier* sold by a newsagent were 276.4 and 12.2 respectively. During 24 days following an advertising campaign, the total number of copies of *The Daily Courier* sold by the newsagent was 6713. Stating your assumptions, test at the 5% significance level whether the data indicate that the campaign was successful.

5.5   An alternative method of carrying out a hypothesis test

Another way of carrying out a hypothesis test is to calculate the probability that the test statistic takes the observed value (or a more extreme value) and to compare this probability with the significance level. If the probability is less than the significance level then the null hypothesis is rejected. The result is said to be 'significant' at the given significance level. Fig. 5.6 shows that this method will always give the same result as the previous method.

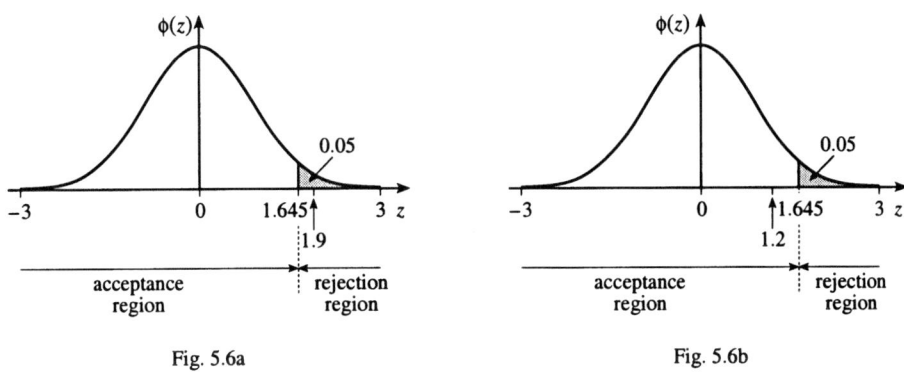

Fig. 5.6a   Fig. 5.6b

Fig. 5.6a shows the rejection region for Z for a one-tail test for an increase at the 5% significance level. If Z takes a value in the rejection region, for example 1.9, then you can see from this figure that $P(Z \geq 1.9)$ is less than 0.05, which would also lead to the rejection of H_0. If Z takes a value in the acceptance region, for example 1.2, as shown in Fig. 5.6b, then $P(Z \geq 1.2)$ is greater than 0.05 and H_0 would not be rejected. To illustrate this idea, look again at Example 5.4.2. In this example, the test statistic took the value -2.572 (correct to 3 decimal places). Low values of Z were of interest so a 'more extreme value' here means a value less than -2.572. The probability that Z took this value or a more extreme value is

$$P(Z \leq -2.572) = 0.005\,06 = 0.506\%, \text{ correct to 3 significant figures.}$$

Since this probability is less than 1%, the result is significant at the 1% level and so the null hypothesis is rejected. As you would expect this is the same as the conclusion which was reached before. However, this way of quoting the result conveys more information: it shows that the result was significant at the 1% level but not at the 0.5% significance level.

Giving a probability (sometimes called a p-value) rather than using critical values requires a little more work. However, with a calculator it is easy to give a probability and it is becoming more common to give the result of a hypothesis test in this form.

The next example illustrates this approach in a two-tail test.

Example 5.5.1
A machine is designed to produce rods 2 cm long with a standard deviation of 0.02 cm. The lengths may be taken as normally distributed. The machine is moved to a new position in the factory, and in order to check whether the setting for the mean length has altered, the lengths of the first ten rods are measured. The standard deviation may be considered to be unchanged. If these lengths, in cm, are as given below, test at the 5% significance level whether the setting has altered or not.

2.04 1.97 1.99 2.03 2.04 2.10 2.01 1.98 1.97 2.02   (OCR, adapted)

This is a two-tail test with the null hypothesis assuming that the mean is unaltered.

The null and alternative hypotheses are $H_0 : \mu = 2$ and $H_1 : \mu \neq 2$ respectively, where μ is the mean length in centimetres of a rod.

Sample mean $= \frac{1}{10}(2.04 + 1.97 + 1.99 + 2.03 + 2.04 + 2.10 + 2.01 + 1.98 + 1.97 + 2.02)$

$\qquad = 2.015$.

Since the population is normally distributed, $\overline{X}$ is also normally distributed.

Under H_0, $\overline{X} \sim N\left(2, \frac{0.02^2}{10}\right)$.

Thus $P(\overline{X} \geq 2.015) = P\left(Z \geq \dfrac{2.015 - 2}{\frac{0.02}{\sqrt{10}}} \right)$

$\qquad = P(Z \geq 2.371...)$

$\qquad = 0.008\,85...$

$\qquad = 0.89\%$, to 2 decimal places.

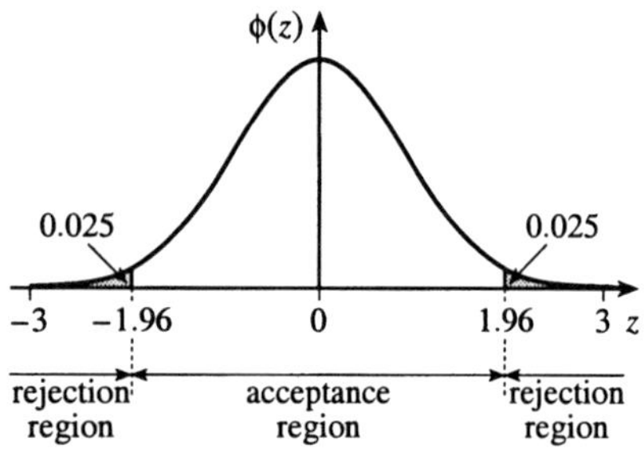

Fig. 5.7

Since this is a two-tail test, this probability should be compared with half of the value specified in the significance level, that is $2\frac{1}{2}\%$, as shown in Fig. 5.7.

Since 0.89% is less than $2\frac{1}{2}\%$, the result is significant at the 5% level. It can be assumed that the mean length of the rods produced by the machine has been affected by the move.

Some calculators give you the choice of either one-tail or two-tail operation. In that case you would end the calculation in this example by finding the two-tailed probability $P(|Z| \geq 2.371...) = 0.0177... = 1.77\%$ and comparing this with the specified significance level of 5%.

Either method of carrying out a hypothesis test, using critical values or using probabilities, is satisfactory and usually you should use the method which you find easier. However, in later chapters you will meet situations where the probability approach is simpler. For this reason, it is suggested that you carry out Exercise 5D using the probability method.

Exercise 5D

Carry out the hypothesis tests in this exercise by calculating probabilities.

1   The random variable X has a normal distribution with mean μ and variance 25. A random sample of 20 observations of X is taken and the sample mean is denoted by $\overline{X}$. This is used to test the null hypothesis $\mu = 30$ against the alternative hypothesis $\mu < 30$.

(a) Calculate $P(\overline{X} \le 28.4)$.

(b) If the sample mean is, in fact, 28.4, state whether the null hypothesis is rejected at the

    (i) 5% significance level,   (ii) 10% significance level.

2   The alkalinity of soil is measured by its pH value. It has been found from many previous measurements that the pH values in a particular area are normally distributed with mean 8.42 and standard deviation 0.74. After an unusually hot summer the pH values were measured at 36 randomly chosen locations in the area and the sample mean value was found to be 8.63. Calculate $P(\overline{X} > 8.63)$ when $\mu = 8.42$.

What can be concluded

(a) at the 5% significance level,   (b) at the 1% significance level?

3   The average time that I have to wait for the 8.15 bus is 4.3 minutes. A new operator takes over the service, with the same timetable, and my average waiting time for 10 randomly chosen days under the new operator is 3.4 minutes. Assuming that the waiting time has a normal distribution with standard deviation 1.8 minutes, test whether the average waiting time under the new operator has decreased. Use a 10% significance level.

4   The marks of all candidates in an A-Level Statistics examination are normally distributed with mean 42.3 and standard deviation 11.2. The 15 candidates entered from Erehwon High School have a mean mark of 49.8. Test, at the 1% significance level, whether Erehwon High School has unusually good results for this examination.

6 Large sample distributions

Chapter 4 described the effect of drawing samples from a population which is known to be normal. This chapter and the next investigate what happens when samples are drawn from populations which are not normal. When you have completed this chapter, you should

- know that, for large samples, the sum of values of the random variable has a distribution which is approximately normal
- know how to interpret a discrete probability in terms of a continuous random variable
- be able to apply this to find normal approximations to binomial probability distributions.

6.1 The probability distribution for a sample total

Example 6.1.1
A dice has the form of a regular icosahedron, with 20 faces. There are faces with 1, 2, 3, 4 and 5 spots, four faces for each number. Find the probability distribution of the total score when the dice is rolled
(a) once,   (b) twice,   (c) four times.

(a)   The probability of getting any one of the scores 1, 2, 3, 4 or 5 is $\frac{4}{20} = 0.2$, so the scores have the probability distribution in Table 6.1.

Score	1	2	3	4	5
Probability	0.2	0.2	0.2	0.2	0.2

Table 6.1

(b)   With two rolls the smallest score you can get is $1 + 1 = 2$ and the largest is $5 + 5 = 10$.

The probabilities of getting the various scores between 2 and 10 can be found from the two-way table in Fig. 6.2. Above the heavy lines to the left and right are the possible scores on the first and second roll respectively, with their probabilities. Below the heavy lines are nine rows of probabilities, expressed as sums, giving the probabilities of getting a total of 2, 3, ... , 10 when the dice is rolled twice.

For example, the third row shows the probability of getting a total of 4. This can happen in three ways: 3 on the first roll and 1 on the second, 2 on the first and 2 on the second, 1 on the first and

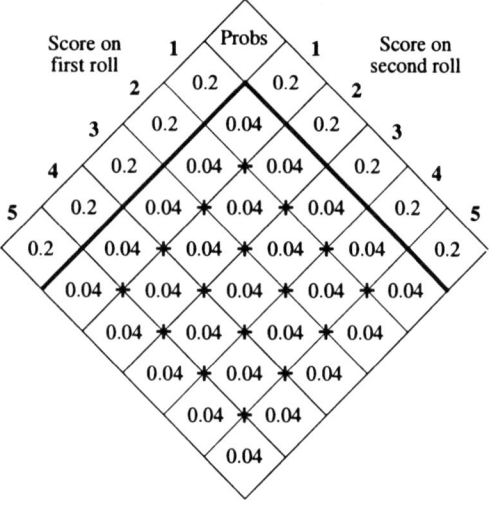

Fig. 6.2

3 on the second. Each of these outcomes has a probability of $0.2 \times 0.2 = 0.04$, so the probability of getting a total of 4 is

$$0.04 + 0.04 + 0.04 = 0.12.$$

Similar calculations for each of the totals produce the probability distribution in Table 6.3.

Total score	2	3	4	5	6	7	8	9	10
Probability	0.04	0.08	0.12	0.16	0.20	0.16	0.12	0.08	0.04

Table 6.3

(c)   With four rolls the smallest score you can get is 4 and the largest is 20. To find the probabilities you can again use a two-way table, Fig. 6.4. Above the heavy line to the left you have the total scores for the first two rolls, with the probabilities assigned to them in Table 6.3. Above the heavy line to the right you have the total scores for the third and fourth rolls, with the same probability distribution. The rows below the heavy lines are sums giving the probabilities of getting a total of 4, 5, ... when the dice is rolled four times. (In Fig. 6.4 only the first seven rows are shown. It is left to you to complete the table and calculate the remaining probabilities.)

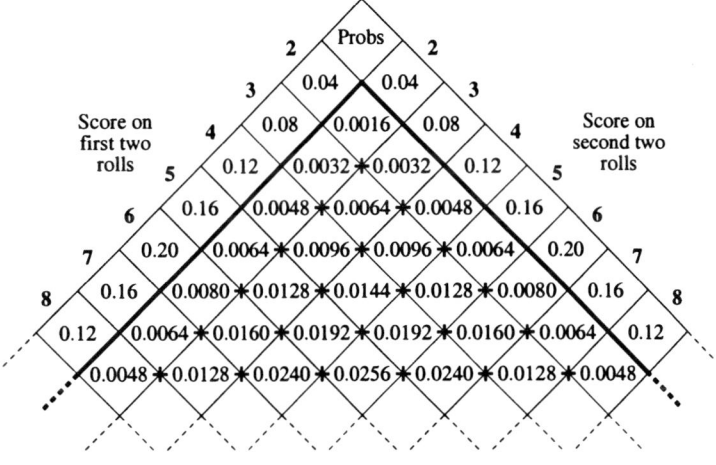

Fig. 6.4

For example, the third row shows the probability of getting a total of 6 with four rolls. This can happen in three ways: 4 on the first two rolls and 2 on the second two, 3 on the first two and 3 on the second two, 2 on the first two and 4 on the second two. From Table 6.3 these outcomes have probabilities of $0.12 \times 0.04 = 0.0048$, $0.08 \times 0.08 = 0.0064$ and $0.04 \times 0.12 = 0.0048$ respectively. So the probability of getting a total of 6 with four rolls is

$$0.0048 + 0.0064 + 0.0048 = 0.0160.$$

Similar calculations for the other rows produce the probability distribution in Table 6.5.

Total score	4	5	6	7	8	9
Probability	0.0016	0.0064	0.0160	0.0320	0.0560	0.0832

Total score	10	11	12	13	14	15
Probability	0.1088	0.1280	0.1360	0.1280	0.1088	0.0832

Total score	16	17	18	19	20
Probability	0.0560	0.0320	0.0160	0.0064	0.0016

Table 6.5

Example 6.1.2

Repeat Example 6.1.1 for an icosahedral dice having eight faces with 2 spots, six with 3 spots, four with 4 spots and two with 5 spots.

(a)   The probabilities of scores of 2, 3, 4, 5 are $\frac{8}{20}$, $\frac{6}{20}$, $\frac{4}{20}$, $\frac{2}{20}$, giving the probability distribution in Table 6.6.

Score	2	3	4	5
Probability	0.4	0.3	0.2	0.1

Table 6.6

(b)   With two rolls the smallest total score is $2 + 2 = 4$ and the largest is $5 + 5 = 10$. The probabilities can be calculated from Fig. 6.7, in the same way as before, except that the probabilities of different scores for each roll are no longer equal. This gives the probability distribution in Table 6.8.

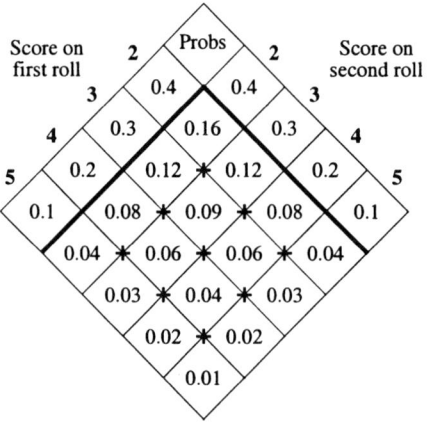

Fig. 6.7

Total score	4	5	6	7	8	9	10
Probability	0.16	0.24	0.25	0.20	0.10	0.04	0.01

Table 6.8

(c)   Proceeding as before, using Fig. 6.9 (which, like Fig. 6.4, is left for you to complete), produces the probability distribution in Table 6.10.

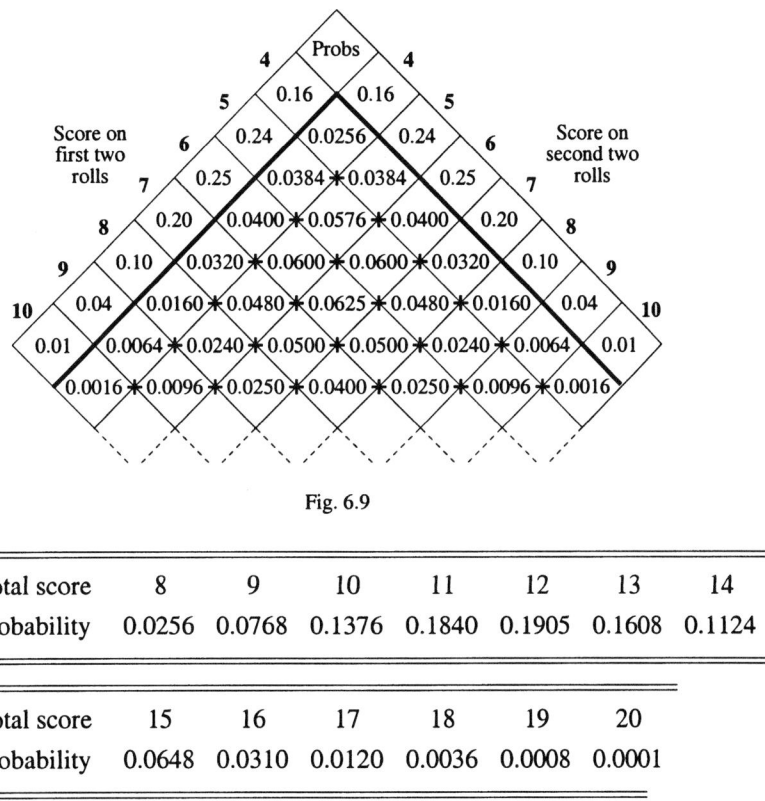

Fig. 6.9

Total score	8	9	10	11	12	13	14
Probability	0.0256	0.0768	0.1376	0.1840	0.1905	0.1608	0.1124

Total score	15	16	17	18	19	20
Probability	0.0648	0.0310	0.0120	0.0036	0.0008	0.0001

Table 6.10

The point of these examples becomes clear when you show these distributions graphically. Figs. 6.11 and Fig. 6.12 show the probability distributions for the dice in Examples 6.1.1 and 6.1.2 respectively.

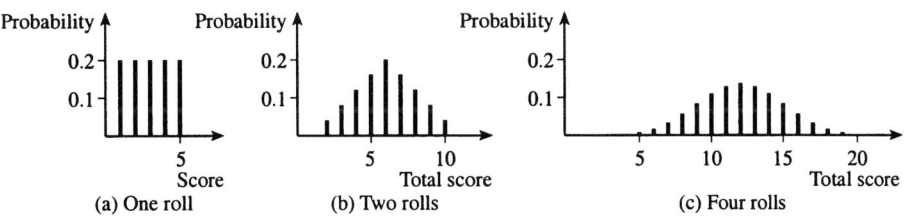

Fig. 6.11

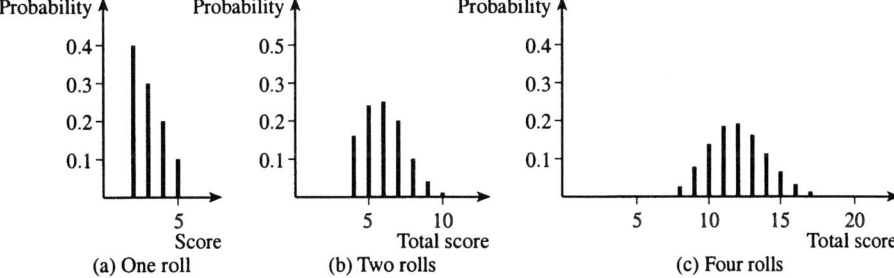

Fig. 6.12

In Fig. 6.11 all the graphs are symmetrical. With one roll the graph is uniform, with two rolls it is triangular, and with four rolls the graph takes a form which will remind you of that of normal probability density.

The graphs in Fig. 6.12 are more surprising, because the graph for one roll is unsymmetrical, with positive skew. However, even with two rolls the mode of the graph has begun to move towards the centre, and with four rolls the graph looks quite symmetrical, and has again begun to resemble the shape of the normal probability graph.

What this illustrates is that the sum of n independent values of a random variable approximates to a normal random variable as the value of n increases. In the examples above, even with four rolls the graphs have begun to take the characteristic bell shape, although the approximation to the actual normal equation is not very close. But if you rolled the dice 100 times, the total scores would approximate very closely to a normal distribution.

From Tables 6.1, 6.3 and 6.5 for Example 6.1.1, and Tables 6.6, 6.8 and 6.10 for Example 6.1.2, you can find the means and variances for the various distributions.

This is a good opportunity to check that you know how to do this using your calculator.

The results are summarised in Tables 6.13 and 6.14 for Examples 6.1.1 and 6.1.2 respectively.

Number of rolls	Mean	Variance
1	3	2
2	6	4
4	12	8

Table 6.13

Number of rolls	Mean	Variance
1	3	1
2	6	2
4	12	4

Table 6.14

There is an obvious pattern here: in both examples the mean and variance are both proportional to the number of rolls. This is what you should expect, since it follows directly by applying the rules of expectation algebra in Section 3.2. If you roll the dice n times, the total score is the sum of the scores in n independent single rolls, each with the probability distribution in Table 6.1 or 6.6. If the distribution of the score X for a single roll has mean μ and variance σ^2, then for the sum of n scores the mean and variance are

$$E(X_1 + X_2 + \ldots + X_n) = E(X_1) + E(X_2) + \ldots + E(X_n)$$
$$= \mu + \mu + \ldots + \mu = n\mu$$

and

$$\mathrm{Var}(X_1 + X_2 + \ldots + X_n) = \mathrm{Var}(X_1) + \mathrm{Var}(X_2) + \ldots + \mathrm{Var}(X_n)$$
$$= \sigma^2 + \sigma^2 + \ldots + \sigma^2 = n\sigma^2.$$

You can put this together with the evidence from Fig. 6.11 and Fig. 6.12 that the graphs approach the form of the normal probability density function to obtain the following result.

> The sum of n independent values of a random variable with mean μ and variance σ^2 has a distribution which, for large values of n, approximates to the normal distribution $N(n\mu, n\sigma^2)$.

This is true whether the random variable is discrete (as in the examples above) or continuous. There is a small complication when you apply it with a discrete random variable, which is discussed in the next section. Here is an example where it is applied to a continuous random variable.

Example 6.1.3

A map of a motorway shows the distances between successive junctions, each correct to the nearest kilometre. A motorist planning a journey calculates the distance between Junction 1 and Junction 13 by adding twelve of these distances. What is the probability that she gets an answer which is correct to the nearest kilometre?

The error in each recorded distance is between $-\frac{1}{2}$ km and $\frac{1}{2}$ km, with uniform probability distribution. The probability density is therefore

$$f(x) = \begin{cases} 1 & \text{for } -\frac{1}{2} < x < \frac{1}{2}, \\ 0 & \text{otherwise.} \end{cases}$$

The mean of this distribution is obviously 0, and the variance is

$$\int_{-\frac{1}{2}}^{\frac{1}{2}} x^2 \times 1\, dx - 0^2 = \left[\tfrac{1}{3}x^3\right]_{-\frac{1}{2}}^{\frac{1}{2}}$$

$$= \tfrac{1}{24} - \left(-\tfrac{1}{24}\right) = \tfrac{1}{12}.$$

So the sum of twelve of these errors has a distribution which is approximately normal, with mean $12 \times 0 = 0$ and variance $12 \times \frac{1}{12} = 1$.

For the motorist to get an answer correct to the nearest kilometre, the total error must lie between $-\frac{1}{2}$ km and $\frac{1}{2}$ km. You therefore need to find the probability that a $N(0,1)$ random variable is between $-\frac{1}{2}$ and $\frac{1}{2}$. That is,

$$P\left(-\tfrac{1}{2} < x < \tfrac{1}{2}\right) = 0.3829\ldots\,.$$

The probability that the answer will be correct to the nearest kilometre is approximately 0.383.

The use of the word 'approximately' in the last sentence of the solution is a reminder of a difficulty in applying the result in the shaded box above. There is no general rule for telling whether 12 is a large enough value of n for the normal approximation to give an answer which is accurate to 3 decimal places. This will depend not just on the value of n but also on the nature of the probability distribution for the original random variable X. The more this original distribution resembles normal probability, the smaller n needs to be to give a reasonable approximation. In this example the original distribution is symmetrical about 0, and Fig. 6.11(c) (for a similar discrete distribution) suggests that the sum of 12 errors is likely to produce a distribution which is quite close to the normal form. So the answer of 0.383 obtained from the normal approximation is probably quite close to the actual probability.

━━━━━━━━━━━━━━━━━ **Exercise 6A** ━━━━━━━━━━━━━━━━━

1   A random variable X has a probability density function

$$f(x) = \begin{cases} \frac{3}{4}\left(1-x^2\right) & \text{for } -1 < x < 1, \\ 0 & \text{otherwise.} \end{cases}$$

Find the probability that the sum of 50 independent values of the random variable is greater than 2.

2   A continuous random variable X has a probability density function

$$f(x) = \begin{cases} \frac{1}{2}x & \text{for } 0 < x < 2, \\ 0 & \text{otherwise.} \end{cases}$$

Find the probability that the sum of 90 independent values of the random variable is between 110 and 130.

3   A firm of caterers wishes to buy wine for a wedding reception of 200 guests. They estimate that, on average, each guest will drink 45 cl of wine. The volume of wine in the bottles they buy may be assumed to have a distribution with mean 70.5 cl and standard deviation 1.2 cl. Show that if they buy 128 bottles then the caterers can be more than 95% certain that their requirements will be met.

4   A shoe repairer knows from experience that the mean time that it takes him to complete a job is 25 minutes, with standard deviation 8 minutes. Find the probability that he can get 100 jobs done inside a 40-hour week.

5   An electronic device emits bleeps independently at random at an average rate of one every 10 seconds. State an approximate normal probability model for the time that will elapse before the hundredth bleep occurs. Calculate the probability that there will be at least 100 bleeps in 15 minutes.

6.2 Interpreting discrete probability in continuous terms

You may have thought it rather odd to describe the graphs in Fig. 6.11(c) and Fig. 6.12(c) as 'approximately normal'. These are graphs of discrete distributions, represented by vertical line segments at each integer value of the random variable. How can they resemble the continuous curve which represents normal probability density?

The answer is to represent the discrete distribution in a slightly different way, by a bar chart that looks similar to a histogram. You widen each of the vertical line segments by a $\frac{1}{2}$ unit on either side, so that it becomes a rectangular bar of width 1 unit centred on an integer value. Fig. 6.15 shows this applied to the discrete probability graph in Fig. 6.12(c).

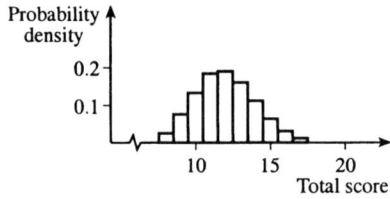

Fig. 6.15

What this does is in effect to replace a discrete random variable X defined by

$$P(X = i) = p_i \quad \text{for } i \in \mathbb{Z}$$

by a continuous random variable X' with probability density defined by

$$f(x') = p_i \quad \text{for } i - \tfrac{1}{2} \le x' < i + \tfrac{1}{2}.$$

It is this random variable X' which is approximated by a normal random variable.

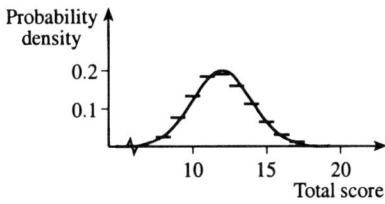

In Fig. 6.16 the graphs of the random variable X' corresponding to Fig. 6.12(c) and the normal probability distribution $N(12, 4)$ are shown with the same axes. You can see that, even with just four rolls of the dice, the two graphs are close to each other.

Fig. 6.16

You could object that the graph of X' is made up of horizontal line segments, so it could never really be said to approximate to the smooth normal curve. But in practice what is important about probability density is not the graph itself but the area under the graph, which represents the probability that the random variable lies within a given interval. You would probably agree that if, in Fig. 6.16, you take an interval of values of the random variable, then the areas under the two graphs over this interval would be close to each other.

6.3 A continuity correction

When you use the normal approximation to calculate probabilities in a discrete distribution, you need to take care in choosing the bounds of the interval for which the area under the normal curve is calculated.

Suppose, for example, that for the dice in Example 6.1.2 you want to find the probability of getting a total of 10, 11 or 12 with four rolls. You can, of course, find the exact answer from Table 6.10, as

$$0.1376 + 0.1840 + 0.1905 = 0.5121.$$

What would you get if you used the normal approximation?

In Fig. 6.15, the exact probability is represented by the sum of the areas of the bars extending from 9.5 to 10.5, from 10.5 to 11.5, and from 11.5 to 12.5. So the corresponding area under the normal curve for $N(12, 4)$ must be taken over the interval $9.5 \le x \le 12.5$. The calculator gives this as 0.4931.

You might think that 0.4931 is not a very close approximation to the exact value of 0.5121. But considering how skewed the probability distribution is for a single dice, and that four is hardly a 'large number', the agreement between the normal approximation and the exact value is in fact remarkably good.

Example 6.3.1

A boy simulates the results of football matches by rolling a dice. If he throws a 1, 2 or 3, his team wins (3 points). If he throws a 4, the result is a draw (1 point). If he throws a 5 or 6, his team loses (0 points). What is the probability that in this simulation his team will get at least 60 points in a season of 30 matches?

The probability distribution of X, the number of points the boy's team scores in a single match, is shown in Table 6.17.

Points	0	1	3
Probability	$\frac{1}{3}$	$\frac{1}{6}$	$\frac{1}{2}$

Table 6.17

So $E(X) = 0 \times \frac{1}{3} + 1 \times \frac{1}{6} + 3 \times \frac{1}{2} = \frac{5}{3}$

and $\text{Var}(X) = \left(0^2 \times \frac{1}{3} + 1^2 \times \frac{1}{6} + 3^2 \times \frac{1}{2}\right) - \left(\frac{5}{3}\right)^2 = \frac{17}{9}$.

The sum, S, of 30 independent values of X has a distribution which is approximately $N\left(30 \times \frac{5}{3}, 30 \times \frac{17}{9}\right)$, which is $N\left(50, \frac{170}{3}\right)$.

If you represented the exact probabilities of the various values of S by a bar chart, then the bar for 60 points would extend from 59.5 to 60.5, the bar for 61 points from 60.5 to 61.5, and so on. So all the bars for 'at least 60 points' would extend from 59.5 upwards.

Using the normal approximation, the probability of getting at least 60 points is therefore given by $P(S \geq 59.5)$. Using a calculator, with $\mu = 50$ and $\sigma = \sqrt{\frac{170}{3}}$, this probability is $0.1034\ldots$.

The probability that the team will get at least 60 points in the season is approximately 0.103.

The interpretation of 'the probability of at least 60 points' as $P(S \geq 59.5)$ is a natural consequence of replacing the discrete random variable taking only integer values by a continuous random variable. It is called a **continuity correction**. Although the probability 0.103 found in the example is still an approximation to the actual probability, it is considerably closer to the actual value than you would get by calculating $P(S \geq 60)$ with $\mu = 50$ and $\sigma = \sqrt{\frac{170}{3}}$, which gives the value 0.092.

Exercise 6B

1   A dice has the form of a regular icosahedron, with 20 faces. There are faces with 1, 2, 3, 4 and 5 spots, four faces for each number. This dice is rolled four times. Find the probability of getting a total score between 10 and 15 (inclusive)

   (a) exactly,   (b) using the normal approximation.

2   A certain brand of matches has the statement 'average contents 45 matches' printed on the box . In fact, boxes contain 43, 44, 45, 46, 47 matches with probabilities 0.1, 0.2, 0.4, 0.2, 0.1. A customer buys a packet of 30 boxes of these matches. Calculate the probabilities that he gets

   (a) exactly 1350 matches,   (b) between 1345 and 1355 matches (inclusive).

3 A shopkeeper decides to get rid of small change by rounding all payments at the check-out to the nearest 10 cents. If the amount ends in 5 cents, he gives the customer the benefit. On one day he serves 800 customers. Assuming that the numbers of 'odd cents' are randomly distributed, find the probability that on that day he will lose more than 5 dollars as a result of this policy.

4 Two children play games of scissors/paper/stone. (Scissors cut paper, paper wraps stone, stone blunts scissors.) In each game they make their choice randomly, and the outcome is given in the following table.

		B chooses		
		scissors	paper	stone
A chooses	scissors	draw	A wins	B wins
	paper	B wins	draw	A wins
	stone	A wins	B wins	draw

Find the probability that, after 60 games, one player leads the other by more than 10 wins.

5 A table of random digits (between 0 and 9) is arranged in 50 rows and 80 columns. Find the expectation of the number of columns in which the sum of the 50 random digits is less than 200.

6.4 A normal approximation to binomial probability

Example 6.4.1
In a city election there is 35% support for the Happy Party candidate, but only 3% support for the Angry Party candidate. A researcher selects 80 citizens at random and asks them how they propose to vote. Draw graphs (in the form of bar charts) showing the probability distributions of the number of people interviewed who support each candidate.

At each interview the researcher will register 1 Happy supporter with probability 0.35, and 0 with probability 0.65. The total number of pro-Happy responses is equal to the number of 1s she registers in all 80 interviews. This therefore has binomial probability with $n = 80$ and $p = 0.35$, denoted by $B(80, 0.35)$. Its graph is shown in Fig. 6.18.

The probability distribution for the total number of pro-Angry supporters is $B(80, 0.03)$. Its graph is shown in Fig. 6.19.

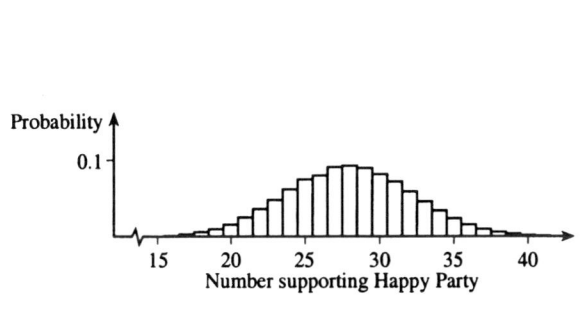

Fig. 6.18

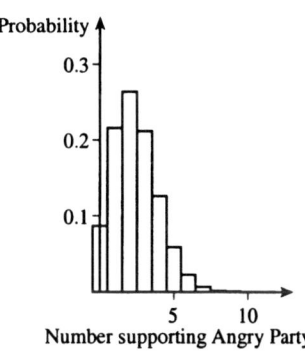

Fig. 6.19

The graphs in Example 6.4.1 look very different. You would probably agree that Fig. 6.18 looks very close to a normal probability graph, but Fig. 6.19 less so. The reason for this is that, for each interview, the probabilities of a 'no' and a 'yes' for the Happy Party are not too unbalanced (0.65 as against 0.35), whereas for the Angry Party the balance is heavily weighted on the 'no' side (0.97 as against 0.03).

In general, binomial probability results from a sequence of separate independent trials, for each of which there are just two possible outcomes – success and failure. A single trial of this kind is called a **Bernoulli trial**. If the random variable X denotes the number of successes in a single trial, then X can take values 0 or 1; and if the probability of success is p and the probability of failure is q (where $q = 1 - p$), the probability distribution for X is given in Table 6.20.

X	0	1
Probability	q	p

Table 6.20

This is the **Bernoulli probability distribution**. It is denoted by $B(1, p)$, because it is simply a binomial distribution $B(n, p)$ with $n = 1$.

Jacob Bernoulli, after whom the distribution is named, was the oldest of a remarkable family of mathematicians which extended over three generations through the late 17th and the 18th century.

From Table 6.20 you can calculate

$$\mu = 0 \times q + 1 \times p = p,$$

and

$$\sigma^2 = \left(0^2 \times q + 1^2 \times p\right) - p^2$$
$$= p - p^2$$
$$= p(1 - p) = pq.$$

> The mean and variance of the Bernoulli probability distribution $B(1, p)$ are
>
> $$\mu = p, \quad \sigma^2 = pq.$$

The general binomial probability distribution gives the probability of the number of successes in a sequence of n independent Bernoulli trials with the same value of p. You can then apply the result in the shaded box in Section 6.1, and deduce a normal distribution to which the binomial approximation approximates when n is large. This normal distribution has mean and variance $n\mu$ and $n\sigma^2$, that is np and npq.

> For large values of n the binomial probability distribution $B(n, p)$ approximates to normal probability density $N(np, npq)$.

For example, the distribution in Fig. 6.18 approximates to $N(80 \times 0.35, 80 \times 0.35 \times 0.65)$, which is $N(28, 18.2)$.

But Example 6.4.1 shows that the validity of this approximation doesn't only depend on the value of n. Both Fig. 6.18 and Fig. 6.19 were drawn with $n = 80$, but you have seen that Fig. 6.18 approximates closely to a normal distribution, but Fig. 6.19 doesn't. A useful working rule is that, for a good approximation, the values of np and nq should both exceed 10.

> The approximation $B(n, p) \approx N(np, npq)$ can be used with confidence if $np > 10$ and $nq > 10$.

That is, binomial probability can be reliably approximated by normal probability if the expected numbers of successes and failures are greater than 10.

Some statisticians have a less demanding criterion, and use the normal approximation if the expected numbers of successes and failures are greater than 5.

Thus for Fig. 6.18, $np = 80 \times 0.35 = 28$ and $nq = 80 \times 0.65 = 52$, both of which are greater than 10, so the normal approximation is valid. But for Fig. 6.19, $np = 80 \times 0.03 = 2.4$ which is less than 10, so the graph is not approximately normal.

Example 6.4.2
It is estimated that 20% of men take size 8 shoes, and 2% take size 12. A shoe store expects 120 male customers in the next week, and buys in

(a)   30 pairs of size 8,   (b)   3 pairs of size 12.

Without using a cumulative binomial program, calculate the probabilities that these will be enough.

(a)   If X denotes the number of size 8 customers, this has probability distribution $B(120, 0.2)$, and you want to find the probability that X is less than or equal to 30.

Since $np = 120 \times 0.2 = 24$ and $nq = 120 \times 0.8 = 96$, and both of these are greater than 10, you can approximate to $B(120, 0.2)$ with the normal probability distribution $N(120 \times 0.2, 120 \times 0.2 \times 0.8)$, which is $N(24, 19.2)$.

The requirement that the discrete random variable X should be less than or equal to 30 is replaced by the requirement that the continuous random variable X' should be less than 30.5. Using the normal approximation, the calculator gives this probability as 0.931, correct to 3 significant figures. Alternatively, using tables of standardised normal probability,

$$P(X' < 30.5) = P\left(Z < \frac{30.5 - 24}{\sqrt{19.2}} \right)$$
$$= P(Z < 1.483\ldots) = 0.931.$$

The probability that the store will have enough size 8 shoes is approximately 0.931.

Compare this with the more accurate answer 0.928 given by the cumulative binomial program.

(b)   The probability distribution for the number Y of size 12 customers is $B(120,0.02)$, for which $np = 120 \times 0.02 = 2.4$. The normal approximation therefore can't be used, and you must calculate the probability that $Y \leq 3$ directly from the binomial formula. That is,

$$P(Y \leq 3) = P(Y = 0) + P(Y = 1) + P(Y = 2) + P(Y = 3)$$

$$= 0.98^{120} + \binom{120}{1} \times 0.98^{119} \times 0.02 + \binom{120}{2} \times 0.98^{118} \times 0.02^2$$

$$+ \binom{120}{3} \times 0.98^{117} \times 0.02^3$$

$$= 0.0885\ldots + 0.2168\ldots + 0.2632\ldots + 0.2113\ldots$$

$$\approx 0.780.$$

The probability that the store will have enough size 12 shoes is 0.780, correct to 3 significant figures.

Example 6.4.3
(a)   Wine is packed in cases, each holding 12 bottles. Because of a fault at the bottling plant, 10% of the bottles contain less wine than is stated on the label. Find the probability that a case contains two or more underfilled bottles.

(b)   The vineyard sends 500 cases to a wine merchant. What is the probability that more than 150 of these will contain two or more underfilled bottles?

   (a)   This is a straight binomial calculation. The distribution of underfilled bottles in a case is $B(12,0.1)$. The calculator program for cumulative binomial probability gives the probability of one or fewer underfilled bottles as 0.6590, correct to 4 decimal places. So the probability of two or more underfilled bottles in a case is $1 - 0.6590 = 0.3410$.

   (b)   The distribution of these cases in a consignment of 500 is $B(500,0.3410)$, which is approximately $N(500 \times 0.3410, 500 \times 0.3410 \times 0.6590)$, which is $N(170.5,112.36)$. In terms of a continuous random variable, 'more than 150 cases' is interpreted as 'greater than 150.5'. The calculator program for normal probability gives the probability as 0.970.

   There is a 97% probability that more than 150 of the cases will contain at least two underfilled bottles.

Exercise 6C

1   A random variable, X, has a binomial distribution with parameters $n = 40$ and $p = 0.3$. Use a suitable approximation, which you should show is valid, to calculate the following probabilities.
   (a) $P(X \geq 18)$ (b) $P(X < 9)$ (c) $P(X = 15)$ (d) $P(11 < X < 15)$

2   The mass production of a cheap pen results in there being 1 defective pen in 20 on average. Use an approximation, which you should show is valid, to find, in a batch of 300 of these pens, the probability of there being
   (a)   24 or more defective pens,   (b)   10 or fewer defective pens.

3 A manufacturer states that '3 out of 4 people prefer our product (Acme) to a competitor's product'. To test this claim a researcher asks 80 people about their liking for Acme. Assuming that the manufacturer is correct, find the probability that fewer than 53 prefer Acme. If 1000 researchers each questioned 80 people, how many of these researchers would be expected to record 'fewer than 53 prefer Acme' results?

4 An ordinary unbiased dice is thrown 900 times. Using a suitable approximation, find the probability of obtaining at least 160 sixes.   (OCR)

5 It is given that 40% of the population support the Gamboge Party. One hundred and fifty members of the population are selected at random. Use a suitable approximation to find the probability that more than 55 out of the 150 support the Gamboge Party.   (OCR)

7 The central limit theorem

It is shown in this chapter that the means of large samples from any population have an approximately normal distribution. When you have completed it, you should

- know and be able to apply the central limit theorem
- know how to apply a continuity correction for the means of samples of a discrete random variable.

7.1 The sampling distribution of the mean

It was shown in Section 4.2 that if you take a sample of n values from a normal population with distribution $N(\mu,\sigma^2)$, then the distribution of the mean of the sample is $N\left(\mu,\dfrac{\sigma^2}{n}\right)$.

But how do you know that the population from which you are taking the sample has a normal distribution?

The practical answer to this question is that, if you take a large enough sample, it doesn't matter.

In Section 6.1 you found that the sum, S, of n values of a random variable taken from *any* population with mean μ and variance σ^2 has a distribution which, for large enough values of n, approximates to the normal distribution $N(n\mu,n\sigma^2)$. If the mean of the sample is $\overline{X}$, then $\overline{X} = \dfrac{1}{n}\times S$. So

$$\mathrm{E}(\overline{X}) = \frac{1}{n}\mathrm{E}(S) = \frac{1}{n}\times n\mu = \mu \quad \text{and} \quad \mathrm{Var}(\overline{X}) = \frac{1}{n^2}\mathrm{Var}(S) = \frac{1}{n^2}\times n\sigma^2 = \frac{\sigma^2}{n}.$$

It follows that, for large values of n, the sample mean $\overline{X}$ has a distribution which approximates to the normal distribution $N\left(\mu,\dfrac{\sigma^2}{n}\right)$. This is called the **central limit theorem.**

> **The central limit theorem**   For any sequence of independent identically distributed random variables $X_1, X_2, \ldots, X_n$ with mean μ and non-zero variance σ^2, provided that n is sufficiently large, $\overline{X} = \dfrac{X_1 + X_2 + \ldots + X_n}{n}$ has approximately a normal distribution with mean μ and variance $\dfrac{\sigma^2}{n}$.
>
> In symbols, $\overline{X} \sim N\left(\mu,\dfrac{\sigma^2}{n}\right)$.

This theorem is a fundamental result in the theory of statistics and it explains why the normal distribution is so widely studied. The essential point is that it does not matter what distribution $X_1, X_2, \ldots, X_n$ have individually: as long as they all have the same distribution and are independent of one another, the distribution of the mean $\overline{X}$ will be approximately normal as long as n is sufficiently large. So the theory in Chapter 4 about the distribution of the sample mean and its application in Chapter 5 to hypothesis testing can be used whether or not the background population is normal, provided that you take a large enough sample.

Example 7.1.1

A continuous random variable, X, has a probability density function, $f(x)$, given by

$$f(x) = \begin{cases} 2x & \text{for } 0 \le x \le 1, \\ 0 & \text{otherwise.} \end{cases}$$

Find   (a) the mean, μ,   (b) the variance, σ^2, of this distribution.

A random sample of 100 observations is taken from this distribution, and the mean $\overline{X}$ is found.

(c) Find the mean and variance of $\overline{X}$.   (d) Calculate $P(\overline{X} < 0.68)$.

(a)   Using the definition of the mean of a continuous random variable,

$$\mu = \int_0^1 x \times 2x \, dx$$

$$= \int_0^1 2x^2 \, dx$$

$$= \left[\tfrac{2}{3} x^3 \right]_0^1 = \tfrac{2}{3}.$$

(b) $$\sigma^2 = \int_0^1 x^2 \times 2x \, dx - \left(\tfrac{2}{3} \right)^2$$

$$= \int_0^1 2x^3 \, dx - \left(\tfrac{2}{3} \right)^2$$

$$= \left[\tfrac{2}{4} x^4 \right]_0^1 - \left(\tfrac{2}{3} \right)^2 = \tfrac{1}{2} - \tfrac{4}{9} = \tfrac{1}{18}.$$

(c) $$\mathrm{E}(\overline{X}) = \mu = \tfrac{2}{3} \quad \text{and} \quad \mathrm{Var}(\overline{X}) = \frac{\sigma^2}{n} = \frac{\tfrac{1}{18}}{100} = \tfrac{1}{1800}.$$

(d)   By the central limit theorem, the distribution of $\overline{X}$ is approximately $\mathrm{N}\!\left(\tfrac{2}{3}, \tfrac{1}{1800} \right)$.

Using a calculator, or by using tables to calculate $P\!\left(Z < \dfrac{0.68 - \tfrac{2}{3}}{\sqrt{\tfrac{1}{1800}}} \right)$, you find that

$P(X < 0.68) = 0.714$, correct to 3 significant figures.

7.2* Applying a continuity correction

When you use the central limit theorem with a discrete population you should apply a continuity correction similar to that used for sums in Section 6.3.

If the discrete random variable takes consecutive integer values, then for sums this involves an adjustment of $\pm \tfrac{1}{2}$ (the $\pm$ sign depending on whether you are dealing with a $>$ or a $<$ condition) to interpret the distribution of sums in terms of a continuous random variable. For means, since these are found by dividing sums by n, the amount of the adjustment is $\pm\dfrac{1}{2n}$.

This is illustrated in Example 7.2.1 for the case when the background population has Bernoulli probability.

Example 7.2.1

About 1 in 7 of the words in the English language begin with the letter S. What is the probability that, in an article of 200 words, fewer than 10% begin with an S?

If there are R occurrences of words beginning with the letter S, then the proportion of these in the article is $\dfrac{R}{200}$. This is the mean number of 'successes' (that is, words beginning with S) in a sequence of 200 Bernoulli trials. The condition 'fewer than 10%' means 'fewer than 20 words', which you would interpret in continuous terms as $R < 19.5$. So 'fewer than 10%' would be interpreted as a proportion of less than $\dfrac{19.5}{200}$, which in continuous terms is 'less than 9.75%'.

For a single Bernoulli trial, with $p = \frac{1}{7}$, the mean is $p = \frac{1}{7}$ and the variance is $pq = \frac{1}{7} \times \frac{6}{7} = \frac{6}{49}$. So, by the central limit theorem, the mean for 200 words has approximately normal distribution $N\left(\dfrac{1}{7}, \dfrac{\frac{6}{49}}{200}\right)$, which is $N\left(\frac{1}{7}, \frac{3}{4900}\right)$.

Using the cumulative normal distribution, the probability that the mean is less than 0.0975 is
$$P\left(Z < \frac{0.0975 - \frac{1}{7}}{\sqrt{\frac{3}{4900}}}\right),$$
which is 0.0334, correct to 3 significant figures.

In this example, notice that the continuity correction involves replacing a proportion of 0.1 by 0.0975, a difference of $0.0025 = \frac{1}{400}$. With $n = 200$, this difference is $\dfrac{1}{2n}$.

Example 7.2.2

(a) If X denotes the number of sixes obtained when a fair cubical dice is thrown 12 times, determine $E(X)$ and $Var(X)$.

(b) Forty students each threw a fair cubical dice 12 times. Each student then recorded the number of times that a six occurred in their own 12 throws. The students' lecturer then calculated the mean number of sixes obtained per student. Give an approximate distribution for this mean and find the probability that it exceeds 2.2.

(a) X satisfies the conditions for a binomial distribution to apply. The parameters of the binomial distribution in this case are $n = 12$ and $p = \frac{1}{6}$.

Recall that for a random variable X which has a binomial distribution, $\mu = np$ and $\sigma^2 = npq$.

So, in this case,

$$\text{E}(X) = 12 \times \tfrac{1}{6} = 2 \quad \text{and} \quad \text{Var}(X) = 12 \times \tfrac{1}{6} \times \tfrac{5}{6} = \tfrac{5}{3}.$$

(b) The sample of students' results is drawn from a background population which is binomial, and which has a graph similar to the normal probability graph. So a sample of size 40 is certainly large enough for the normal approximation to give accurate answers. Using the central limit theorem, $\overline{X} \sim \text{N}\left(2, \dfrac{\tfrac{5}{3}}{40}\right) = \text{N}\left(2, \tfrac{1}{24}\right)$, approximately.

To find the probability that $\overline{X} > 2.2$ you need to apply a continuity correction. Recall that the lecturer calculates $\overline{X}$ by dividing T by 40, where T is the total number of sixes of all the students. So the condition $\overline{X} > 2.2$ is equivalent to $T > 40 \times 2.2 = 88$.

To apply the normal approximation T is replaced by a continuous variable T', and the condition $T > 88$ is replaced by $T' > 88\tfrac{1}{2}$. In the same way, $\overline{X}$ is replaced by a continuous random variable $\overline{X}'$, and you require the probability that $\overline{X}' > 2.2 + \tfrac{1}{80} = 2.2125$.

Notice that, as in Example 7.2.1, the continuity correction is $\dfrac{1}{2n}$, this time with $n = 40$.

With the usual standardisation,

$$\text{P}\left(\overline{X}' > 2.2 + \tfrac{1}{80}\right) = \text{P}\left(Z > \frac{2.2125 - 2}{\sqrt{\tfrac{1}{24}}}\right)$$
$$= 0.149, \text{ correct to 3 significant figures.}$$

When a discrete distribution is approximated by a continuous distribution, the continuity correction for the sample mean, $\overline{X}$, is $\dfrac{1}{2n}$ for a sample of size n.

Exercise 7

1 A random variable X has mean 50 and variance 1000. A random sample of 40 observations of X is taken and the mean, $\overline{X}$, of these observations is calculated. How, approximately, is $\overline{X}$ distributed? Find

(a) $\text{P}(\overline{X} < 55)$,   (b) $\text{P}(\overline{X} > 40)$,   (c) $\text{P}(40 < \overline{X} < 55)$.

2 A discrete random variable, Y, has the probability distribution shown below.

y	1	2	3	4
$P(Y = y)$	0.1	0.2	0.5	0.2

For this distribution, find

(a) the mean, μ,   (b) the variance, σ^2.

A random sample of 50 observations of Y is taken. Find

(c) $P(\overline{Y} < 2.6)$ (d) $P(|\overline{Y} - \mu| < 0.2)$.

3 An unbiased dice is thrown once. Write down the probability distribution of the score X and show that $\mathrm{Var}(X) = \frac{35}{12}$.

The same dice is thrown 70 times.

(a) Find the probability that the mean score is less than 3.3.

(b) Find the probability that the mean exceeds 3.8.

4 A rectangular field is gridded into squares of side $1\,\mathrm{m}$. At one time of the year the number of snails in the field can be modelled by a Poisson distribution with mean 2.25 per m^2.

(a) A random sample of 120 squares is observed and the number of snails in each square counted. Find the probability that the sample mean number of snails is at most 2.5.

(b) A random sample of 100 squares is observed and the number of snails in each square is counted. Find the probability that the sample mean number of snails is at least 2.

5 The random variable X has a $B(40, 0.3)$ distribution. The mean of a random sample of n observations of X is denoted by $\overline{X}$. Find

(a) $P(\overline{X} \geq 13)$, assuming a sample size of 49,

(b) the smallest value of n for which $P(\overline{X} \geq 13) < 0.001$.

6 The number of tickets sold each day at a city railway station during the winter has mean 512 and variance 1600. For a randomly chosen period of 60 winter days, find the probability that the mean number of tickets sold per day over this period is less than 500.

8 Hypothesis testing with discrete variables

This chapter takes further the idea of hypothesis testing introduced in Chapter 5. When you have completed it, you should

- be able to formulate hypotheses and carry out a hypothesis test of a population proportion by either evaluation of binomial probabilities or by the use of the normal approximation
- understand the difference between a nominal and an actual significance level.

8.1 Testing a population proportion

Advertisements for dairy spreads have claimed that the spread cannot be distinguished from butter. How could you set about testing this claim? One way would be to take pairs of biscuits and put butter on one biscuit in each pair and the dairy spread on the other. The pairs of biscuits would be given to a number of tasters who would be asked to identify the biscuit with butter on it. Half the tasters would be given the 'butter' biscuit first, and the other half the 'butter' biscuit second.

Suppose you decided to use 10 tasters. How would you set about drawing a conclusion from your results? The method of hypothesis testing described in Chapter 5 can be adapted to this situation. First it is necessary to formulate a null hypothesis and an alternative hypothesis. It is usual to start from a position of doubt: you assume that the tasters cannot identify the butter and that they are guessing. In this situation the probability that a taster chosen at random will get the correct result is $\frac{1}{2}$. This can be expressed by the null hypothesis $H_0: p = \frac{1}{2}$. If some of the tasters can actually identify the butter then $p > \frac{1}{2}$. This can be expressed as an alternative hypothesis, $H_1: p > \frac{1}{2}$.

Can you see why it is difficult to take any other null hypothesis?

If H_0 is true, the number, X, of tasters who identify the buttered biscuit correctly is a random variable with distribution $B\left(10, \frac{1}{2}\right)$. Fig. 8.1 shows this distribution.

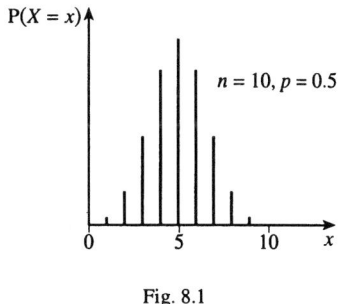

Fig. 8.1

High values of X would suggest that H_0 should be rejected in favour of H_1. The most straightforward method of carrying out a hypothesis test for a discrete variable is to use the approach of Section 5.5 and calculate the probability that X takes the observed or a more extreme value assuming that H_0 is true and compare this probability with the specified significance level. Suppose that 9 out of the 10 people had identified the butter and you chose a significance level of 5%. Using the cumulative binomial probability program on your calculator gives

$$= 1 - 0.989\,25\ldots = 0.010\,74\ldots = 1.07\%, \text{correct to 3 significant figures.}$$

This probability is less than 5% so the result is significant at the 5% level. H_0 is rejected and there is evidence, at the 5% significance level, that the proportion of people who can distinguish the butter is greater than $\frac{1}{2}$.

As in Section 5.3 the possible values of X can be divided into an acceptance region and a rejection region. However the situation here is complicated by the fact that X is a discrete variable. Table 8.2 shows the probability distribution of X.

x	0	1	2	3	4	5	6	7	8	9	10
$P(X = x)$	0.0010	0.0098	0.0439	0.1172	0.2051	0.2461	0.2051	0.1172	0.0439	0.0098	0.0010

Table 8.2

For a nominal significance level of, say, 5%, you will find that there is no rejection region which exactly corresponds to this probability. For example, for a rejection region of $X \geq 8$, the probability of a result in the rejection region is $0.0439 + 0.0098 + 0.0010 = 0.0547$ and so the actual significance level of the test is 5.47%; for $X \geq 9$, the actual significance level is 1.07%. This point will be considered in more detail in Section 9.3.

Example 8.1.1

A national opinion poll claims that 40% of the electorate would vote for party R if there were an election tomorrow. A student at a large college suspects that the proportion of young people who would vote for them is lower. She asks 16 fellow students, chosen at random from the college roll, which party they would vote for. Three choose party R. Show, at the 10% significance level, that this indicates that the reported figure is too high for the young people at the student's college.

The null and alternative hypotheses are $H_0: p = 0.4$ and $H_1: p < 0.4$ respectively.

Let X be the number of students who choose party R. Under H_0, $X \sim B(16, 0.4)$.

Using the cumulative binomial probability program on your calculator,

$P(X \leq 3) = 0.0651\ldots = 6.51\%$, correct to 3 significant figures.

This probability is less than 10% so the result is significant at the 10% level. H_0 is rejected, indicating that the reported figure is too high for young people at the student's college.

Example 8.1.2

In order to test a coin for bias it is tossed 20 times. The result is 14 heads and 6 tails. Test, at the 10% significance level, whether the coin is biased.

This is a two-tail test since, before the coin is tossed, there is no indication in which direction, if any, it might be biased. If the coin is unbiased, the probability of a head (or a tail) is $p = 0.5$.

The null and alternative hypotheses are $H_0: p = 0.5$ and $H_1: p \neq 0.5$ respectively.

Let X be the number of heads resulting from 20 tosses. Under H_0, $X \sim B(20, 0.5)$. On average you would expect 10 heads, so the observed value of 14 is on the high side. Using your calculator,

$P(X \geq 14) = 1 - P(X \leq 13)$

$= 1 - 0.942\,34\ldots = 0.057\,65\ldots = 5.77\%$, correct to 3 significant figures.

Since this is a two-tail test at the 10% significance level, this probability must be compared with 5%. Since 5.77% > 5% the result is not significant and the null hypothesis is not rejected: there is insufficient evidence, at the 10% level, to say that the coin is biased.

To carry out a hypothesis test on a discrete variable, calculate the probability of the observed or a more extreme value and compare this probability with the significance level. For a one-tail test, reject the null hypothesis if this probability is less than the significance level; for a two-tail test use the side that produces a probability of less than 50% and reject the null hypothesis if this probability is less than half of the significance level.

Exercise 8A

1 A large housing estate contains a children's playground, and on one particular evening 12 boys and 6 girls were playing there. Assuming these children are a random sample of all children living on the estate, test, at the 10% significance level, whether there are equal numbers of boys and girls on the estate.

2 An advertisement in a newspaper inserted by a car dealer claimed, 'At least 95% of our customers are satisfied with our services.'

In order to check this statement a random sample of 25 of the dealer's customers were contacted and 22 agreed that they were satisfied with the dealer's services. Carry out a test, at the 5% significance level, of whether the data support the claim.

3 The lengths of nails produced by a machine have a normal distribution with mean 2.5 cm. A random sample of 16 nails is selected from a drum containing a large number of these nails. The nails are measured and 13 are found to have length greater than 2.5 cm. Test, at the $2\frac{1}{2}$% significance level, whether the mean length of the nails in the drum is greater than 2.5 cm. State where in the test the information that the nails have a normal distribution is used.

4 A test of 'telepathy' is devised using cards with faces coloured either red, green, blue or yellow, in equal numbers. When a card is placed face down on a table, at random, Ilesh believes he can forecast the colour of the face correctly. The cards are thoroughly mixed, one is selected and placed face down on the table and Ilesh forecasts the colour of the face. This procedure was repeated 20 times.

(a) It is given that Ilesh was correct on 8 occasions. Test, at the 5% significance level, whether Ilesh's results were better than could have been achieved by chance.

(b) How many forecasts would Ilesh have had to make to be significant at the 1% level?

5 A dice is suspected of being loaded to give more sixes when thrown than would be expected from a fair dice. In order to test this suspicion, the dice is thrown 30 times. Ten sixes are obtained. Carry out a test at the 5% significance level to test this suspicion.

6 A magazine article reported that 80% of computer owners use the internet facility regularly. Lisa believed that the true figure was different and she consulted 12 of her friends who owned computers. Six said that they were regular users of the internet facility.

(a) Test Lisa's belief at the 10% significance level.

(b) Comment on the reliability of the test in the light of Lisa's sample.

8.2   Testing a population proportion for large samples

When the sample is large an alternative method of calculating probabilities is to use the fact that the binomial distribution can be approximated by the normal distribution with a continuity correction. This is especially useful in problems for which you need to use the inverse function, because your calculator has an inverse program for normal probability but no corresponding program for binomial probability.

Example 8.2.1

In a multiple choice paper with 100 questions a candidate has to select one of four possible answers to each question. How many questions would a candidate need to answer for the examiner to be persuaded, at the 5% significance level, that he is not guessing the answers?

If the student is guessing the answers, then the probability p that any one answer is correct is $\frac{1}{4}$. If the student is not guessing then the proportion of correct answers should be greater than this.

The null and alternative hypotheses are $H_0: p = \frac{1}{4}$ and $H_1: p > \frac{1}{4}$.

Let X be the number of correct answers. Under H_0, $X \sim B\left(100, \frac{1}{4}\right)$. This can be approximated by a normal distribution with

$$\mu = np = 100 \times \tfrac{1}{4} = 25 \quad \text{and} \quad \sigma^2 = npq = 100 \times \tfrac{1}{4} \times \tfrac{3}{4} = 18.75.$$

So $X \sim B\left(100, \frac{1}{4}\right)$ is approximated by $V \sim N(25, 18.75)$.

Suppose that the student needs to get k answers correct to persuade the examiner that he is not guessing. Then using a significance level of 5% you need to find k such that

$$P(X \geq k) < 0.05.$$

With a continuity correction the condition $X \geq k$ is equivalent to $V > k - 0.5$. So in terms of the normal approximation you want to find an integer k such that

$$P(V > k - 0.5) < 0.05.$$

For a one-tail test at the 5% significance level the rejection region for the test statistic $Z \sim N(0,1)$ is $Z \geq 1.645$ (Section 5.4). So k has to satisfy the inequality

$$\frac{(k - 0.5) - 25}{\sqrt{18.75}} \geq 1.645,$$

that is $k \geq 32.6$. Since k has to be an integer, this means that the smallest number of correct answers the student must get is 33.

You can use the cumulative binomial program for $B\left(100, \frac{1}{4}\right)$ to check this.

$$P(X \geq 32) = 1 - P(X \leq 31)$$
$$= 1 - 0.9306\ldots = 0.069$$

and

$$P(X \geq 33) = 1 - P(X \leq 32)$$
$$= 1 - 0.9554\ldots = 0.045,$$

correct to 3 decimal places. This confirms that the smallest value of k such that $P(X \geq k) < 0.05$ is 33.

Example 8.2.2

If births are equally likely on any day of the week, the proportion of babies born at the weekend should be $\frac{2}{7}$. A researcher suspects that this is not the case, and decides to test it from the records of the 490 births in the city during a recent month. If she uses a two-tail test at the 5% significance level, what numbers of weekend births would confirm her suspicions?

If p is the probability of a birth occurring at a weekend, the null and alternative hypotheses would be $H_0 : p = \frac{2}{7}$ and $H_1 : p \neq \frac{2}{7}$.

Let X denote the number of babies born at the weekend. Under H_0, $X \sim B\left(490, \frac{2}{7}\right)$. This can be approximated by $V \sim N\left(490 \times \frac{2}{7}, 490 \times \frac{2}{7} \times \frac{5}{7}\right) = N(140, 100)$.

For a two-tail test at the 5% significance level the rejection region is $|Z| > 1.960$, so you would reject values of V such that $\left| \dfrac{V - 140}{\sqrt{100}} \right| > 1.960$; that is, $V < 120.4$ or $V > 159.6$. With a continuity correction, these values correspond to

$$X < 120.4 - 0.5 = 119.9 \quad \text{or} \quad X > 159.6 + 0.5 = 160.1.$$

But X must be an integer, so this gives a rejection region of $X \leq 119$ or $X \geq 161$.

However, since the method involved an approximation, these answers should be checked using the cumulative binomial probabilities. These give

$$P(X \leq 119) = 0.0189 \ldots, \quad P(X \leq 120) = 0.0243 \ldots, \quad P(X \leq 121) = 0.0308 \ldots,$$
$$P(X \geq 160) = 0.0267 \ldots, \quad P(X \geq 161) = 0.0212 \ldots.$$

This shows that, since $P(X \leq 120) < 0.025$, $X = 120$ should also be included in the rejection region.

So the researcher's suspicions would be confirmed if the number of weekend births was 120 or less, or 161 or more.

Exercise 8B

1   A jar contains a large number of coloured beads. It is claimed that 30% of them are red. A random sample of 80 beads is selected. What numbers of red beads in the sample would be required to support the claim at the 10% significance level?

2   The manufacturers of a new cold relief drug believe that more than 75% of people suffering from a cold will find it effective. They test it with 150 volunteers. How many effective responses would be needed to confirm the manufacturers' belief at the $2\frac{1}{2}\%$ significance level?

3   A parcel delivery service claims that at least 80% of their parcels are delivered within 48 hours of posting. A check was carried out on 200 parcels. What is the smallest number that should be delivered within 48 hours for the claim to be accepted, using a 5% significance level?

4   The drop-out rate of students enrolled at a certain university is reported to be 13.2%. The Dean of Students suspects that the drop-out rate for science students is greater than 13.2%, and she examines the record of a random sample of 95 of these students. How many drop-outs would the Dean need to find in the sample if her suspicions were to be confirmed at the 2% significance level?

9 Errors in hypothesis testing

This chapter investigates the situation where the wrong conclusion is drawn from a hypothesis test. When you have completed it, you should

- know what Type I and Type II errors are
- be able to calculate Type I and Type II errors in the context of the normal and binomial distributions.

9.1 Type I and Type II errors

When you carry out a hypothesis test your final step is to reject or to accept the null hypothesis. For example, the teachers described in Section 5.1, who were trying out a new reading scheme, had to choose between

(a) the new reading scheme is better than the old one, or
(b) the new reading scheme is not better than the old one.

If they came to conclusion (a) they would probably introduce the new scheme; if they came to conclusion (b) they would probably stick to their current reading scheme.

When such a decision is made after carrying out a hypothesis test, it may be either correct or incorrect. You can never be absolutely certain that you have made the right decision because you have to rely on a limited amount of evidence. For example, the reading scheme can only be tested on a sample of children. The situation is similar to that in a trial where the defendant is found either guilty or not guilty on the basis of the evidence brought forward. In this case there are four possible situations (which are mutually exclusive).

The defendant is innocent and is found not guilty: in this case the decision is correct.
The defendant is innocent but is found guilty: in this case the decision is incorrect.
The defendant is guilty but is found not guilty: in this case the decision is incorrect.
The defendant is guilty and is found guilty: in this case the decision is correct.

Suppose that in a criminal court of law a defendant is assumed innocent unless found guilty 'beyond reasonable doubt'. The initial assumption of innocence is equivalent to the null hypothesis, and the theory that the defendant is guilty is equivalent to the alternative hypothesis. Deciding what constitutes 'beyond reasonable doubt' is equivalent to setting a significance level. Similarly in a hypothesis test there are four possible situations, again mutually exclusive.

H_0 is true and H_0 is accepted: in this case the decision is correct.
H_0 is true but H_0 is rejected: in this case the decision is incorrect.
H_0 is not true but H_0 is accepted: in this case the decision is incorrect.
H_0 is not true and H_0 is rejected: in this case the decision is correct.

You can see that there are two different ways in which an incorrect decision could be made. In order to distinguish between them they are called Type I and Type II errors.

> A **Type I error** is made when a true null hypothesis is rejected.
>
> A **Type II error** is made when a false null hypothesis is accepted.

Making an incorrect decision can be costly in various ways. For example, suppose that a fire alarm was tested to see whether it was still functioning correctly after a power cut. You might take as the null and alternative hypotheses

H_0: the alarm is functioning correctly,
H_1: the alarm is not functioning correctly.

A Type II error in this situation would mean that you assumed that the alarm was functioning correctly when in fact it was not. This could result in injury, loss of life or damage to property. A Type I error would mean that you thought the alarm was not working correctly when in fact it was. This could mean expenditure on unnecessary repairs or replacement.

Try to analyse in a similar way the 'costs' of making Type I and Type II errors for the reading scheme example.

The examples given in this chapter should make you appreciate that it is important to assess the risk of making errors when carrying out a hypothesis test. In order to do this you have to calculate

$$P(\text{Type I error}) = P(\text{rejecting } H_0 \mid H_0 \text{ true})$$
and $\quad P(\text{Type II error}) = P(\text{accepting } H_0 \mid H_0 \text{ false}).$

The following sections show you how these probabilities are calculated for the different types of test which you have met in earlier chapters.

9.2   Type I and Type II errors for tests involving the normal distribution

If you look back to Section 5.3, which considered continuous variables, you will see that the probability of the test statistic falling in the rejection region, when H_0 is true, is equal to the significance level of the test. If the test statistic falls in the rejection region then H_0 will be rejected when it is in fact true; that is, a Type I error will be made.

> When the distribution of the test statistic is continuous, P(Type I error) is equal to the significance level of the test.

The choice of significance level for a hypothesis test is thus related to the value of P(Type I error) which you are prepared to accept. The choice of a significance level should depend in the first instance on how serious the consequences of a Type I error are. The more serious the consequences, the lower the value of the significance level which should be used. For example, if the consequences of a Type I error were not serious, you might use a significance level of 10%; if the consequences were very serious you might use a significance level of 0.1%.

Consider the reading scheme example. A Type I error in this case would mean that the new scheme was adopted even though it did not produce better results. As a result money would be wasted on a new scheme which was no better than the old. If the new scheme is not any better and the teachers use a 5% significance level then there is a 1 in 20 chance that the money will be wasted. If the new material is very costly then the teachers might feel that such a risk is unacceptable and choose a significance level of 1% or even less depending on the resources of their school. If on the other hand the new scheme is not very expensive, or they need to replace their reading material anyway, then they might take a significance level of 10% or even 20%.

A Type II error involves accepting a false null hypothesis, which means that you fail to detect a difference in μ. You would expect the probability of this happening to depend on how much μ has changed: if there is a small difference in μ it could easily go undetected but if there is a big difference in μ then you would expect to detect it. This is why the alternative hypothesis has to be defined more exactly before P(Type II error) can be calculated.

The following example illustrates the method.

Example 9.2.1
A machine fills 'one litre' cola bottles. When the machine is working correctly the contents of the bottles are normally distributed with mean 1.002 litres and standard deviation 0.002 litre. The performance of the machine is tested at regular intervals by taking a sample of 9 bottles and calculating their mean content. If this mean content falls below a certain value, it is assumed that the machine is not performing correctly and it is stopped.

(a) Set up null and alternative hypotheses for a test of whether the machine is working correctly.

(b) For a test at the 5% significance level, find the rejection region taking the sample mean as the test statistic.

(c) Give the value for the probability of making a Type I error.

(d) Find P(Type II error) if the mean content of the bottles has fallen to the nominal value of 1.000 litre.

(e) Find the range of values of μ for which the probability of making a Type II error is less than 0.001.

(a) $H_0: \mu = 1.002$ (the machine is working correctly);

$H_1: \mu < 1.002$ (the mean contents has fallen).

(b)   Under H_0, $\overline{X} \sim N\left(1.002, \dfrac{0.002^2}{9}\right)$.

For a one-tail test for a decrease at the 5% level, the rejection region for the test statistic Z is $Z \le -1.645$. Since $Z = \dfrac{\overline{X} - \mu}{\frac{\sigma}{\sqrt{n}}}$ this means that $\dfrac{\overline{X} - 1.002}{\frac{0.002}{\sqrt{9}}} \le -1.645$.

Rearranging gives the rejection region for the sample mean as $\overline{X} \le 1.000\,90$.

In Fig. 9.1, the broken curve shows the distribution of $\overline{X}$ if H_0 is true and the hatched area shows the P(Type I error).

(c)   For a continuous test statistic, P(Type I error) = significance level = 0.05.

(d) $P(\text{Type II error}) = P(\text{accepting } H_0 \mid H_0 \text{ false})$

$$= P(\overline{X} > 1.000\ 90 \mid \mu = 1.000),$$

that is $P(\overline{X}$ is in acceptance region $\mid \mu$ is no longer 1.002 but 1.000).

This probability is shown by the solid shaded area in Fig. 9.1, where the solid curve shows the distribution of $\overline{X}$ if H_1 is true.

$$P(\overline{X} > 1.0009 \mid \mu = 1.000) = P\left(Z > \frac{1.000\ 90 - 1.000}{\frac{0.002}{\sqrt{9}}} \right)$$

$$= P(Z > 1.35)$$

$$= 0.088, \text{ correct to 3 decimal places.}$$

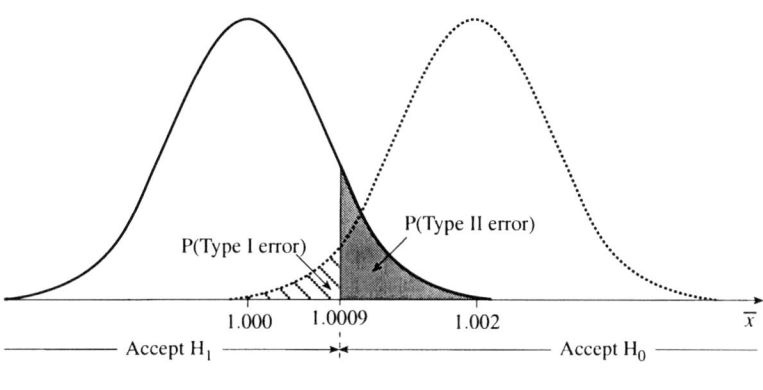

Fig. 9.1

(e)   First find the value of z for which the probability of making a Type II error is 0.001. Looking back to part (d) this would require $P(Z > z) = 0.001$. Using the inverse normal distribution in tables or on your calculator, $z = 3.090$.

A Type II error will be made if the test statistic Z is greater than 3.0902, that is if

$$\frac{1.000\ 90 - \mu}{\frac{0.002}{\sqrt{9}}} > 3.0902.$$

Solving gives $\mu < 0.9988$, correct to 4 decimal places.

So for $\mu < 0.9988$ the probability of making a Type II error is less than 0.001.

In Example 9.2.1 the consequence of setting the significance level at 5% is that there is a probability of 5% of stopping the machine unnecessarily when it is working correctly. With this significance level the probability of failing to detect that the mean contents of the bottles has fallen to the nominal value of 1.000 litre is 8.77%. It is interesting to see what happens to P(Type II error) when a lower significance level is used. This is done in the next example.

Example 9.2.2

Repeat Example 9.2.1 parts (b) to (d) with a significance level of 1%.

(b) For a one-tail test for a decrease at the 1% level, the rejection region for the test statistic Z is $Z \leq -2.326$ so

$$\frac{\overline{X} - 1.002}{\frac{0.002}{\sqrt{9}}} \leq -2.326.$$

Rearranging gives the rejection region for the sample mean as $\overline{X} \leq 1.000\,45$.

(c) $P(\text{Type I error}) = \text{significance level} = 0.01$.

(d) $P(\text{Type II error}) = P(\text{accepting } H_0 \mid H_0 \text{ false})$
$$= P(\overline{X} > 1.000\,45 \mid \mu = 1.000),$$

that is $P(\overline{X}$ is in acceptance region $\mid \mu$ is no longer 1.002 but 1.000$)$.

$$P(\overline{X} > 1.000\,4\ldots \mid \mu = 1.000) = P\left(Z > \frac{1.000\,45 - 1.000}{\frac{0.002}{\sqrt{9}}} \right)$$
$$= P(Z > 0.675)$$
$$= 0.250, \text{ correct to 3 decimal places.}$$

For this second calculation, at the 1% significance level, the value of $P(\text{Type I error})$ has been reduced but the value of $P(\text{Type II error})$ has increased. You would expect this result from looking at Fig. 9.1. If the critical value is altered so that one type of error increases, the other will decrease. This means that in setting a significance level it may be necessary to assess the risks involved in committing both types of error and balance one against the other. The only way in which both types of error can be reduced at the same time is by taking a larger sample, so that the overlap of the distributions in Fig. 9.1 is reduced.

The following example shows how to calculate $P(\text{Type II error})$ for a two-tail test.

Example 9.2.3

Boxes of a certain breakfast cereal have contents whose masses are normally distributed with mean μ and standard deviation 15 grams. A test of the null hypothesis $\mu = 375$ against the alternative hypothesis $\mu \neq 375$ is carried out at the 5% significance level using a random sample of 16 boxes.

(a) For what values of the sample mean is the alternative hypothesis accepted?

(b) Given that the actual value of μ is 380, find the probability of making a Type II error.

(OCR, adapted)

(a) Under H_0, $\overline{X} \sim N\left(375, \frac{15^2}{16}\right)$.

For a two-tail test at the 5% significance level, the rejection region is $|Z| \geq 1.960$.

Now $Z = \dfrac{\overline{X} - 375}{\dfrac{15}{\sqrt{16}}}$.

You can check that when $Z = 1.96$, $\overline{X} = 382.35$, and when $Z = -1.96$, $\overline{X} = 367.65$. Thus the alternative hypothesis is accepted when $\overline{X} \geq 382.35$ or $\overline{X} \leq 367.65$.

(b) $P(\text{Type II error}) = P(367.65 < \overline{X} < 382.35 \mid \mu = 380)$

$$= P\left(\frac{367.65 - 380}{\dfrac{15}{\sqrt{16}}} < Z < \frac{382.35 - 380}{\dfrac{15}{\sqrt{16}}} \right)$$

$$= P(-3.293\ldots < Z < 0.626\ldots)$$

$$= 0.734, \text{ correct to 3 significant figures.}$$

Exercise 9A

1   The random variable X has a normal distribution with mean μ and variance 12.8. A test, at the 5% significance level, of the null hypothesis $\mu = 5$ against the alternative hypothesis $\mu > 5$ is carried out using a random sample of 20 observations of X.

(a)   Give the rejection region of the test, in terms of the sample mean, $\overline{X}$.

(b)   Find the probability of a Type II error in the test when the true value of μ is 7.

2   In a test of the quality of Luxiglow paint, which is intended to cover an area of at least 10 m^2 per litre can, a random sample of 15 cans is tested. The mean area per can covered by the 15 cans is denoted by $\overline{X} \text{ m}^2$. It may be assumed that the area covered by a can has a normal distribution with standard deviation 0.51 m^2.

(a)   Find, in terms of $\overline{X}$, the rejection region of a test, at the $2\tfrac{1}{2}\%$ significance level, that the mean area covered by all litre cans of the paint is at least 10 m^2.

(b)   For a particular sample, $\overline{X} = 10.3$. State the type of error that could not occur.

(c)   Given that the mean cover per can of paint is actually 9.6 m^2, calculate the probability of making a Type II error in the test.

3   In a quality control check 5 randomly selected packs of butter are weighed. The masses of all packs of butter may be assumed to have a normal distribution with mean μ grams and standard deviation 2.7 grams. A test of the null hypothesis $\mu = 247$ against the alternative hypothesis $\mu \neq 247$ is carried out at the $\alpha\%$ significance level. It is decided to accept the null hypothesis if the sample mean lies between 245 grams and 249 grams.

(a)   Find the value of α.

(b)   Given that the actual value of μ is 250, find the probability of making a Type II error in the test.

(c)   What can be said about the probability of making a Type II error when the value of μ is greater than 250?

4   The number of daily absences by employees of a large company has mean 1.94 and standard deviation 0.22. A new system of working is introduced in the hope that this will reduce the number of absences, and it is found that there were 68 absences during the first 40 days of the new system. Treating the 40 days as a random sample

   (a) test, at the 5% significance level, whether the new system had the desired effect,

   (b) calculate the probability of making a Type II error in the test in part (a) when the mean number of absences is actually 1.8,

   (c) state, in the context of the question, what is meant by a Type II error.

5   Studies have shown that the time taken for adults to memorise a list of 12 words has mean 3.8 minutes and standard deviation 1.8 minutes. Taking a course in mnemonics is believed to reduce the mean. To investigate this belief a test, at the 5% significance level, is proposed based on a random sample of 36 people who took the course. Each was given the same list of 12 words to memorise.

   (a) Find the rejection region of the test in terms of the sample mean time. Assume that the standard deviation remains at 1.8 minutes.

   (b) Given that the actual mean time for the 12 words is 2.9 minutes after taking the course, find the probability of making a Type II error in the test.

   Suppose now that the test is based on a random sample of 40 people.

   (c) Show that the probability of making a Type II error (when the actual mean time is 2.9 minutes) is smaller than that found in part (b).

6*   The breaking strength of lengths of wire required in the manufacture of a certain piece of machinery has a normal distribution with mean 30 N and standard deviation 0.38 N. A random sample of 9 lengths of the wire is tested to determine whether the population mean breaking strength is less than 30 N. A Type I error for the test has probability 0.04.

   (a) Find the set of values of the sample mean breaking strength for which it would be accepted that the mean breaking strength is not less than 30 N.

   (b) Given that the probability of making a Type II error in the test is to be less than 0.025, find the set of possible values of the actual mean breaking strength.

9.3   Type I and Type II errors for tests involving the binomial distribution

In Section 8.1 you met the idea that, for a discrete distribution, it is not usually possible to find a rejection region which corresponds exactly to the specified significance level. You may find it helpful to look back at Section 8.1 before going on to the following example.

In this section all binomial probabilities are given correct to 4 decimal places.

Example 9.3.1

An experiment on telepathy is carried out by two people. One person, A, chooses a card at random from a standard pack and concentrates on it. The other person, B, who cannot see the card, has to write down the suit of the card. This is done for 18 cards in all and X, the number of cards whose suit is correctly identified, is counted.

(a)   State suitable hypotheses, involving a probability, for a hypothesis test which could indicate whether person B is able to name the correct suit more often than would be expected by chance.

(b)   What would be the rejection region for the test statistic X for a test at the 10% significance level?

(c)   The nominal significance level of this test is 10%. What is the actual significance level?

(a)   There are four suits, so if person B is guessing, the probability of being correct is $\frac{1}{4}$; if person B has telepathic powers the probability of being correct will be greater than this, so take $H_0: p = \frac{1}{4}$ and $H_1: p > \frac{1}{4}$.

(b)   Under H_0, $X \sim B(18, 0.25)$. Let the rejection region for X be denoted by $X \geq c$. You need to find the smallest value of c for which $P(X \geq c)$ is less than the given significance level. From the cumulative binomial probability program on your calculator you will see that, if H_0 is true,

$$P(X \geq 7 \mid p = 0.25) = 1 - P(X \leq 6 \mid p = 0.25)$$
$$= 1 - 0.8610 = 0.1390 = 13.9\% > 10\%,$$

so H_0 is accepted.

$$P(X \geq 8 \mid p = 0.25) = 1 - P(X \leq 7 \mid p = 0.25)$$
$$= 1 - 0.9431 = 0.0569 = 5.69\% < 10\%,$$

so H_0 is rejected.

Thus, the rejection region is $X \geq 8$.

(c)   The actual significance level of the test is $P(X \geq 8 \mid p = 0.25) = 5.69\%$. This is equal to $P(\text{Type I error})$.

This is not very close to the desired significance level (10% in this example) and this will often be the case in tests involving discrete variables.

> For a hypothesis test involving a discrete variable, for example a variable which has a binomial or Poisson distribution, the rejection region is defined so that
>
> $P(\text{test statistic falls in rejection region} \mid H_0 \text{ true})$
> $\leq$ nominal significance of the test.
>
> Actual significance level of the test =
> $P(\text{test statistic falls in rejection region} \mid H_0 \text{ true})$
>
> and this is also the probability of a Type I error.

The following example shows how to calculate $P(\text{Type II error})$.

Example 9.3.2

A supplier of primrose seeds claims that their germination rate is 0.95. A purchaser of the seeds suspects that the germination rate is lower. To test this claim the purchaser plants 20 seeds in similar conditions, counts the number, X, which germinate and carries out a hypothesis test at the 5% significance level.

(a) Formulate suitable null and alternative hypotheses to test the seed supplier's claim.

(b) For what values of X would the null hypothesis be rejected?

(c) The nominal significance level of this test is 5%. What is the actual significance level?

(d) What is the probability of a Type I error using this test?

(e) Calculate $P(\text{Type II error})$ if the probability that a seed germinates is in fact 0.80.

    (a) $H_0: p = 0.95$, $H_1: p < 0.95$.

    (b)   Low values of X will lead to the null hypothesis being rejected and so the rejection region takes the form $X \le c$. Under H_0, $X \sim B(20, 0.95)$. The cumulative binomial probability program shows that

$$P(X \le 16 \mid p = 0.95) = 0.0159 = 1.59\% < 5\%, \text{ so } H_0 \text{ is rejected;}$$

$$P(X \le 17 \mid p = 0.95) = 0.0755 = 7.55\% > 5\%, \text{ so } H_0 \text{ is accepted.}$$

        So H_0 is rejected if $X \le 16$, and the rejection region for the test is $X \le 16$.

    (c) $P(\text{Type I error}) = P(X \le 16 \mid p = 0.95) = 0.0159$.

    (d)   The actual significance level $= P(\text{test statistic falls in rejection region} \mid H_0 \text{ is true})$
$$= 0.0159 = 1.59\%.$$

    (e) $P(\text{Type II error}) = P(X > 16 \mid p = 0.8)$
$$= 1 - P(X \le 16 \mid p = 0.8)$$
$$= 1 - 0.5886 = 0.4114.$$

The value of P(Type II error) in Example 9.3.2 indicates that there is a fairly high probability that the hypothesis test will fail to detect a fall in the germination rate from 0.95 to 0.80.

The following example illustrates a two-tail test.

Example 9.3.3

The dice used in a board game is rolled 30 times and the number, X, of sixes is counted.

(a) Set up suitable null and alternative hypotheses for testing whether the dice is biased either towards or away from six.

(b) What would be the result of a test at the 10% significance level if $X = 9$?

(c) What is the rejection region for X for a test at the 10% significance level?

(d) What is $P(\text{Type I error})$ for this test?

(e) Find $P(\text{Type II error})$ if the dice is in fact biased so that the probability of getting a six is 0.5.

    (a) $H_0: p = \frac{1}{6}$, $H_1: p \ne \frac{1}{6}$.

(b)   Under H_0, $X \sim B\left(30, \frac{1}{6}\right)$. The value $X = 9$ is on the high side since, on average, you would expect 5 sixes in 30 throws. From the cumulative binomial probability program

$$P\left(X \geq 9 \mid p = \frac{1}{6}\right) = 1 - P\left(X \leq 8 \mid p = \frac{1}{6}\right)$$
$$= 1 - 0.9494 = 0.0506 = 5.06\% > 5\%,$$

so H_0 is not rejected.

The probability is compared with $\frac{1}{2} \times 10\% = 5\%$ since a two-tail test is being carried out.

There is not enough evidence, at the 10% significance level, that the dice is biased.

(c)   The null hypothesis could be rejected for either high or low values of X (since a two-tail test is being carried out) and so there are two parts to the rejection region. Firstly you need to find a value c such that $P\left(X \leq c \mid p = \frac{1}{6}\right) \leq 0.05$. The cumulative binomial probability program shows that

$$P\left(X \leq 2 \mid p = \frac{1}{6}\right) = 0.1028 \geq 0.05, \text{ so accept } H_0,$$

and $P\left(X \leq 1 \mid p = \frac{1}{6}\right) = 0.0295 \leq 0.05,$ so reject H_0.

So the lower part of the rejection region is $X \leq 1$.

Secondly you need a value d such that $P\left(X \geq d \mid p = \frac{1}{6}\right) \leq 0.05$. You found in part (b) that $X = 9$ is not rejected, so try $X = 10$.

$$P\left(X \geq 10 \mid p = \frac{1}{6}\right) = 1 - P\left(X \leq 9 \mid p = \frac{1}{6}\right)$$
$$= 1 - 0.9803 = 0.0197 \leq 0.05.$$

So the upper part of the rejection region is $X \geq 10$.

The null hypothesis is rejected when $X \leq 1$ or $X \geq 10$.

(d) $P(\text{Type I error}) = P\left(X \leq 1 \text{ or } X \geq 10 \mid p = \frac{1}{6}\right)$
$$= 0.0295 + 0.0197 = 0.0492 = 4.92\%.$$

This is well below the desired significance level of 10%. It would be possible to get a significance level closer to this value by taking the rejection region as $X \leq 1$ and $X \geq 9$, which gives a significance level of 8.01%. However, it is usual practice if a two-tail test is required to calculate the rejection region as in part (c) unless you are asked to find a rejection region such that $P(\text{Type I error})$ is as close as possible to the nominal significance level.

(e)   Taking the rejection region found in part (c) gives

$$P(\text{Type II error}) = P(1 < X < 10 \mid p = 0.5)$$
$$= P(X \leq 9 \mid p = 0.5) - P(X \leq 1 \mid p = 0.5)$$
$$= 0.0214 - 0.0000 = 0.0214.$$

In Example 9.3.3 the low value of $P(\text{Type II error})$ indicates that the hypothesis test would be effective in detecting a dice biased so that the probability of getting a six is 0.5.

For large samples it becomes easier to find a rejection region which gives $P(\text{Type I error})$ close to the required significance level because you can use the normal approximation to the binomial distribution (see Section 8.2). The following example shows how to calculate P(Type I error) and P(Type II error) in this situation.

Example 9.3.4

It is suspected that a gaming club is running an unfair roulette wheel. The wheel is given 3700 trial spins and X, the number of times that zero (on which the club wins) turns up is counted. There are 37 possible scores on a trial spin, labelled 0 to 36, and these should have equal probability.

(a) For what values of X would you conclude that the wheel was biased in favour of zero if a hypothesis test was carried out at the 5% significance level?

(b) For this test calculate P(Type II error) if the probability of getting zero is in fact $\frac{3}{37}$.

(OCR, adapted)

(a)   If the wheel is fair then the probability of getting zero is $\frac{1}{37}$. If the wheel is unfair it will favour the club so the probability of getting zero is greater than this.

The null and alternative hypotheses are $H_0: p = \frac{1}{37}$ and $H_1: p > \frac{1}{37}$ respectively.

Under H_0, $X \sim B\left(3700, \frac{1}{37}\right)$.

Using the normal approximation,

$$\mu = np = 3700 \times \tfrac{1}{37} = 100, \quad \text{and} \quad \sigma^2 = npq = 3700 \times \tfrac{1}{37} \times \tfrac{36}{37} = \tfrac{3600}{37},$$

so $X \sim B\left(3700, \frac{1}{37}\right)$ is approximated by $V \sim N\left(100, \frac{3600}{37}\right)$.

For a one-tail test for an increase at the 5% significance level, H_0 is rejected if $Z \geq 1.645$. The corresponding rejection region for V is given by

$$\frac{V - 100}{\sqrt{\frac{3600}{37}}} \geq 1.645 \quad \text{giving} \quad V \geq 116.2, \text{ correct to 1 decimal place.}$$

Allowing for the continuity correction, this corresponds to $X \geq 116.2 + 0.5 = 116.7$.

Rounding up to the nearest integer, this suggests that the rejection region is $X \geq 117$. You can check this by using the cumulative binomial program on your calculator, which gives

$$P(X \geq 116) = 0.0605 > 5\% \quad \text{and} \quad P(X \geq 117) = 0.0498 < 5\%.$$

(b)   If $p = \frac{3}{37}$, then the distribution of X can be approximated by

$$V \sim N\left(3700 \times \tfrac{3}{37}, 3700 \times \tfrac{3}{37} \times \tfrac{34}{37}\right) = N\left(300, \tfrac{10\,200}{37}\right).$$

$$P(\text{Type II error}) = P(X \leq 116)$$
$$\approx P(V \leq 116.5) \quad \text{(including the continuity correction)}$$
$$= P\left(Z < \frac{116.5 - 300}{\sqrt{\frac{10\,200}{37}}}\right) = P(Z < -11.05...) \approx 0.$$

The hypothesis test described in Example 9.3.4 would be extremely effective at detecting a degree of bias such that the probability of getting zero is $\frac{3}{37}$.

Exercise 9B

1 A newspaper reported that 55% of households own more than two television sets. Each of a random sample of 30 households in Melchester is contacted and the number of households owning more than two television sets is denoted by N. A test of whether the proportion p of households in Melchester owning more than two television sets is different from 55% is carried out at a nominal 5% significance level.

    (a) Obtain the rejection region of the test in terms of N.

    (b) Calculate P(Type I error).   (c) State the conclusion of the test when $N = 20$.

    (d) Calculate P(Type II error) when the actual value of p is 60%.

2 An election to the presidency of a society of 8000 members is shortly to take place. After a pre-election speech by Mrs Robinson (a candidate), a random sample of 25 people who listened to the speech were asked about their voting intentions. The number who say that they will vote for Mrs Robinson is denoted by R. A test is carried out at a nominal 5% significance level of whether Mrs Robinson will be elected, which will happen if she gets more than 50% of the votes cast.

    (a) Find the rejection region of the test in terms of R and state the conclusion of the test when $R = 19$.

    (b) What is the actual significance level of the test?

    (c) Given that 65% of all members will vote for Mrs Robinson, find the probability of making a Type II error in the test.

3 A drug for treating phlebitis has proved effective in 75% of cases when it has been used. A new drug has been developed which, it is believed, will be more successful and it is used on a sample of 16 patients with phlebitis. A test is carried out to determine whether the new drug has a greater success rate than 75% and the test statistic is X, the number of patients cured by the new drug. It is decided to accept that the new drug is more effective if $X > 14$.

    (a) Find α, the probability of making a Type I error.

    (b) Find β, the probability of making a Type II error when the actual success rate is 80%.

    What can be said about the values of α and β if, with the same decision procedure $(X > 14)$, the sample size were larger than 16?

4 Of a certain make of electric toaster, 10% have to be returned for service within three months of purchase. A modification to the toaster is made in the hope that it will be more reliable. Out of 24 modified toasters sold in a store none was returned for service within three months of purchase. The proportion of all the modified toasters that are returned for service within three months of purchase is denoted by p.

    (a) State, in terms of p, suitable hypotheses for a test.

    (b) Test whether there is evidence, at a nominal 10% significance level, that the modified toaster is more reliable than the previous model in that it requires less service.

    (c) What is the probability of making a Type I error in the test?

    (d) Assuming that, when n is large, $B(n, p)$ is nearly symmetrical, estimate the set of values of p for which P(Type II error) < 0.25.

10 The *t*-distribution

This chapter explains how you can carry out a hypothesis test when the background population has a normal distribution whose variance is unknown. When you have completed it, you should

- understand the use of S_{n-1}^2 as an unbiased estimator of the population variance
- know how to use the random variable T to describe the distribution of the sample mean
- understand the idea of degrees of freedom
- be able to apply the *t*-distribution to test hypotheses about the value of the population mean.

10.1 Estimating the population variance

In the last few chapters you have worked many problems like this (Exercise 9A Question 2):

> In a test of the quality of Luxiglow paint,… a random sample of 15 cans is tested…. . It may be assumed that the area covered by a can has a normal distribution with standard deviation 0.51 m^2. Find … the rejection region of a test… that the mean area covered by all litre cans of the paint is at least 10 m^2.

It may have struck you as odd that there should be doubt about the value of the mean, but that the standard deviation (or the variance) should be precisely known. In most situations it is more likely that, when you take a sample to test a hypothesis about the mean of a population, you won't know the variance either. How then could you estimate its value?

The obvious answer is to use the variance of the sample as a guide to the variance of the population. You will remember from Higher Level Book 1 Section 7.8 that a so-called 'unbiased estimate' of the variance of a population can be calculated as $\dfrac{n}{n-1} s_n^{\ 2}$, where $s_n^{\ 2}$ is the variance of a sample of n measurements. This quantity is often denoted by $s_{n-1}^{\ 2}$.

When you use a calculator to find the variance of a sample, it is very important to know whether the quantity calculated is $s_n^{\ 2}$ or $s_{n-1}^{\ 2}$.

This section explores this distinction in more detail. The first example looks at the problem the other way round. It begins with a very simple population with known variance, and finds the variance of all the samples of a given size that can be taken from it. The purpose is to compare the variance of the samples with the variance of the population.

Example 10.1.1
A population consists of just two numbers -1 and $+1$. Write down all the samples (with replacement)
(a) of size 2,   (b) of size 3,
that can be drawn from this population. Compare the variance of the samples with the variance of the population.

Begin by noting that the mean of the population is 0, so the variance is $\dfrac{(-1-0)^2 + (1-0)^2}{2} = 1$.

(a)   There are four samples of size 2 which can be drawn from the population:

$$(-1,-1),(-1,+1),(+1,-1),(+1,+1).$$

The variances of these are 0, 1, 1, 0 respectively.

Notice that two of the samples have the same variance as the population, and two have a smaller variance. The mean of the four variances is $\dfrac{0+1+1+0}{4}=\dfrac{1}{2}$.

(b)   There are eight samples of size 3:

$$(-1,-1,-1),(-1,-1,+1),(-1,+1,-1),(-1,+1,+1),$$
$$(+1,-1,-1),(+1,-1,-1),(+1,+1,-1),(+1,+1,+1).$$

The variances are $0,\dfrac{8}{9},\dfrac{8}{9},\dfrac{8}{9},\dfrac{8}{9},\dfrac{8}{9},\dfrac{8}{9},0$.

None of the samples has a variance as large as the variance of the population. The mean of the variances is $\dfrac{2\times0+6\times\frac{8}{9}}{8}=\dfrac{2}{3}$.

You may be interested to continue this example by finding the variance of the 16 samples of size 4. What would you guess for the mean of the 16 variances?

The most obvious feature of this example is that, for both $n=2$ and $n=3$, the variance of any random sample is not a reliable guide to the variance of the population. But the mean of the variance of all the random samples is, in both cases, $\dfrac{n-1}{n}$ times the variance of the population: $\dfrac{1}{2}$ for $n=2$, $\dfrac{2}{3}$ for $n=3$. This is in fact true for samples of any size taken from any population.

> The mean of the variance of all independent random samples of size n taken from a population with variance σ^2 is equal to $\dfrac{n-1}{n}\sigma^2$.

A proof of this is given in the next section.

With a larger sample from a more extensive population it is impractical to list all the possible samples, but you can test out the result in the shaded box by finding the variance of a number of samples and taking the mean. You won't expect to get exactly $\dfrac{n-1}{n}\sigma^2$, but you should usually get an answer reasonably close to this.

Example 10.1.2

The population of the 10 single digit numbers from 0 to 9 has $\sigma^2=8.25$. Take 10 samples, each of 5 digits, from this population by using random numbers from your calculator or from tables, and calculate their variance. Compare the mean of these variances with the value of $\dfrac{n-1}{n}\sigma^2$ with $n=5$.

You may find it interesting to work through this example with your own random digits instead of the ones used here.

By reading across a row of a table of random digits the following samples were generated.

$(1,8,4,3,9)$	$(7,3,7,6,8)$	$(9,7,5,6,5)$	$(6,5,7,9,5)$	$(0,1,8,8,1)$
$(1,8,9,4,8)$	$(0,7,4,2,6)$	$(0,1,1,9,2)$	$(6,7,0,5,1)$	$(9,5,4,3,2)$

For each sample, calculate the variance.

9.20	2.96	2.24	2.24	13.04
9.20	6.56	10.64	7.76	5.84

The mean of these variances is 6.968. This can be compared with the mean over all possible samples, which is $\frac{n}{n-1}\sigma^2 = \frac{4}{5} \times 8.25 = 6.6$.

Reference has already been made in this section to an 'unbiased estimate' of the variance of a population. In statistics the word 'unbiased' has a precise meaning.

> A random variable R is said to be an **unbiased estimator** of a parameter p of a probability distribution if $E(R) = p$.

So the box on the previous page states that S_n^2, the variance of a random sample of size n, is not an unbiased estimator of the population variance σ^2, because $E(S_n^2) = \frac{n-1}{n}\sigma^2 \neq \sigma^2$.

A capital letter S is used in S_n^2 because the variance of a random sample is itself a random variable.

However, if you multiply both sides of this equation by $\frac{n}{n-1}$, you get

$$E\left(\frac{n}{n-1}S_n^2\right) = \frac{n}{n-1}E(S_n^2) = \sigma^2.$$

This means that $\frac{n}{n-1}S_n^2$ is an unbiased estimator of σ^2.

> If S_n^2 is the variance of a random sample of size n drawn from a population with variance σ^2, then $\frac{n}{n-1}S_n^2$ is a unbiased estimator of σ^2.

Don't imagine that this means that, if you take a particular random sample with variance s_n^2, then $\frac{n}{n-1}s_n^2$ will be a good estimate of the population variance. Example 10.1.2 shows this very clearly. If you had picked on the third sample $(9,7,5,6,5)$, this would estimate the population variance as $\frac{5}{4} \times 2.24 = 2.8$; but if you had picked the fifth sample $(0,1,8,8,1)$, you would get an estimate for the population variance of $\frac{5}{4} \times 13.04 = 16.3$. Neither is very close to the actual population variance of 8.25.

For a sample $(x_1, x_2, \ldots, x_n)$ with mean $\bar{x}$, $s_n^{\,2}$ is calculated as

$$s_n^{\,2} = \frac{(x_1 - \bar{x})^2 + (x_2 - \bar{x})^2 + \ldots + (x_n - \bar{x})^2}{n},$$

so

$$\frac{n}{n-1} s_n^{\,2} = \frac{(x_1 - \bar{x})^2 + (x_2 - \bar{x})^2 + \ldots + (x_n - \bar{x})^2}{n-1}.$$

Because this is like $s_n^{\,2}$ but with $n-1$ in the denominator instead of n, the quantity $\frac{n}{n-1} s_n^{\,2}$ is often denoted by the symbol $s_{n-1}^{\,2}$. This is the notation used in this book. But in other books and on some calculators you may also meet other notations such as $\hat{\sigma}^2$ (the circumflex is used to stand for 'unbiased estimator'), $\sigma_{n-1}^{\,2}$, $\sigma_x^{\,2}$ or $s_x^{\,2}$.

Before ending this section, it is worth noting that the mean of a sample is an unbiased estimator of the mean of the population. To prove this, begin with a population in which a random variable X has mean μ, and take from it a random sample $(X_1, X_2, \ldots, X_n)$. The mean of this sample is

$$\bar{X} = \frac{X_1 + X_2 + \ldots + X_n}{n}.$$

You want to find the expected value of $\bar{X}$ as $X_1, X_2, \ldots, X_n$ independently range over the whole population. So find

$$\begin{aligned} \mathrm{E}(\bar{X}) &= \frac{1}{n}\big(\mathrm{E}(X_1) + \mathrm{E}(X_2) + \ldots + \mathrm{E}(X_n)\big) \\ &= \frac{1}{n}\big(\mathrm{E}(X) + \mathrm{E}(X) + \ldots + \mathrm{E}(X)\big) \\ &= \frac{1}{n} \times n\mathrm{E}(X) = \mathrm{E}(X). \end{aligned}$$

And since $\mathrm{E}(X) = \mu$, it follows that

$$\mathrm{E}(\bar{X}) = \mu.$$

$\bar{X}$ is an unbiased estimator of the population mean μ.

10.2* Proof of the 'mean of variances' formula

This section gives a proof of the result in the first shaded box in Section 10.1. You may omit it if you wish.

With the notation at the end of the last section, suppose that the random variable X has variance σ^2. The problem now is to find the expected value of the variance of the random sample $(X_1, X_2, \ldots, X_n)$ as $X_1, X_2, \ldots, X_n$ independently range over the whole population.

It turns out that the proof is much easier if $\mu = 0$. So it pays to begin by introducing a new random variable Y defined by $Y = X - \mu$. Then

$$\begin{aligned} \mathrm{E}(Y) &= \mathrm{E}(X - \mu) \\ &= \mathrm{E}(X) - \mu \\ &= \mu - \mu = 0, \end{aligned}$$

and

$$\begin{aligned} \mathrm{Var}(Y) &= \mathrm{Var}(X - \mu) \\ &= \mathrm{Var}(X) = \sigma^2. \end{aligned}$$

Also, since

$$\begin{aligned} \mathrm{Var}(Y) &= \mathrm{E}\big((Y - 0)^2\big) \\ &= \mathrm{E}(Y^2), \end{aligned}$$

it follows that

$$\mathrm{E}(Y^2) = \sigma^2.$$

So now take a random sample $(Y_1, Y_2 \ldots, Y_n)$ of independent values of Y. If the mean of this sample is $\bar{Y}$, the variance V of the sample can be calculated as

$$V = \mathrm{E}(Y^2) - \bar{Y}^2 = \frac{Y_1^2 + Y_2^2 + \ldots + Y_n^2}{n} - \left(\frac{Y_1 + Y_2 + \ldots + Y_n}{n}\right)^2.$$

You want the expectation of this as the random variables $Y_1, Y_2, \ldots, Y_n$ range over the population.

Look first at the second term in this expression, whose numerator is $(Y_1 + Y_2 + \ldots + Y_n)^2$. When you multiply this out, you get square terms $Y_1^2, Y_2^2, \ldots, Y_n^2$, and product terms like $2Y_1Y_2$. Now it was proved in Section 3.2 that the expectation of the product of two independent random variables is equal to the product of the expectations of the two variables separately. So

$$\mathrm{E}(2Y_1Y_2) = 2\mathrm{E}(Y_1) \times \mathrm{E}(Y_2).$$

And since Y_1 and Y_2 both have the distribution of Y, it follows that $\mathrm{E}(Y_1) = \mathrm{E}(Y) = 0$, and similarly for $\mathrm{E}(Y_2)$. Therefore the expectations of all the product terms are 0. It follows that

$$\mathrm{E}\big((Y_1 + Y_2 + \ldots + Y_n)^2\big) = \mathrm{E}(Y_1^2 + Y_2^2 + \ldots + Y_n^2),$$

so that

$$\begin{aligned} \mathrm{E}(V) &= \frac{1}{n}\mathrm{E}(Y_1^2 + Y_2^2 + \ldots + Y_n^2) - \frac{1}{n^2}\mathrm{E}(Y_1^2 + Y_2^2 + \ldots + Y_n^2) \\ &= \frac{n-1}{n^2}\big(\mathrm{E}(Y_1^2) + \mathrm{E}(Y_2^2) + \ldots + \mathrm{E}(Y_n^2)\big). \end{aligned}$$

Also, for each of the random variables Y_i, $\mathrm{E}(Y_i^2) = \mathrm{E}(Y^2)$. So

$$E(V) = \frac{n-1}{n^2} \times nE(Y^2)$$

$$= \frac{n-1}{n}E(Y^2)$$

$$= \frac{n-1}{n}\sigma^2,$$

using the result $E(Y^2) = \sigma^2$ proved earlier.

10.3   Finding numerical estimates of population parameters

If you have a sample of n numerical values of a random variable, you can now use the results in Section 10.1 to make numerical estimates of the mean and variance of the population from which the sample has been taken. It is important to understand the distinction between the mean and variance of the sample, and the estimated values of the mean and variance of the population.

If the values $x_1, x_2, \ldots, x_n$ represent a whole population, that is all the values of interest, and you wish to calculate the variance of these values, then use

$$s_n^2 = \sum \frac{x^2}{n} - \bar{x}^2, \text{ or its equivalent } \frac{1}{n}\sum(x - \bar{x})^2.$$

If, on the other hand, you are trying to estimate the variance of a larger population from which the values $x_1, x_2, \ldots, x_n$ are a sample, then use

$$s_{n-1}^2 = \frac{n}{n-1}\left(\sum \frac{x^2}{n} - \bar{x}^2\right), \text{ or its equivalent } \frac{1}{n-1}\sum(x - \bar{x})^2.$$

Example 10.3.1

(a) Nine CDs were played and the playing time of each CD was recorded. The times, in minutes, are given below.

49,   56,   55,   68,   61,   57,   61,   52,   63

Find the mean playing time of the nine CDs and the variance of the playing times.

(b) A student was doing a project on the playing times of CDs. She wished to estimate the mean playing time for CDs sold in the UK and she wished also to estimate the variance of playing times of CDs sold in the UK. She took a sample of nine CDs and recorded their playing times. The results are given below.

49,   56,   55,   68,   61,   57,   61,   52,   63

(i) Use the student's data to estimate the mean playing time for CDs sold in the UK.

(ii) Use the student's data to estimate the variance of the playing times of CDs sold in the UK.

The two parts, (a) and (b) look very similar. However in part (a) you are only interested in the playing times of the nine CDs which have been selected. In part (b) you wish to make estimates of population parameters from the sample data that you have been given.

(a) $\bar{x} = \frac{1}{9}(49 + 56 + \ldots + 63) = 58$, so the mean is 58 minutes.

$$\text{Variance} = \sum \frac{x^2}{n} - \bar{x}^2 = \frac{1}{9}\left(49^2 + 56^2 + \ldots + 63^2\right) - 58^2$$
$$= 30.4\ldots,$$

so the variance is 30.4 min^2, correct to 3 significant figures.

(b)   In this part you are estimating the mean and variance of the population.

(i)   The unbiased estimator of the population mean, μ, is the mean of the sample $\bar{x}$, so the estimate of μ will be 58 minutes.

(ii)   To obtain an unbiased estimate of σ^2, you need to use $s_{n-1}^2 = \frac{n}{n-1}\left(\sum \frac{x^2}{n} - \bar{x}^2\right)$.

As $\sum \frac{x^2}{n} - \bar{x}^2$ has already been calculated, to find s_{n-1}^2 all that needs to be done is to multiply $\sum \frac{x^2}{n} - \bar{x}^2$ by $\frac{n}{n-1}$, which in this case is $\frac{9}{8}$.

Therefore s_{n-1}^2, the unbiased estimate of σ^2, is $\frac{9}{8} \times 30.4\ldots = 34.25$.

Example 10.3.2

A fishing crew recorded the masses in kilograms (kg) of 200 fish of a particular species that were caught on their trawler. The results are summarised in Table 10.1. The weights given are mid-class values.

Weight of fish (kg)	0.5	1.25	1.75	2.25	2.75	3.5	4.5	5.5	7.0	10.5
Number of fish in class	21	32	33	24	18	21	16	12	11	12

Table 10.1

Assuming that these fish are a random sample from the population of this species, estimate

(a) the mean mass, in kilograms, of a fish of this species,

(b) the variance of masses of fish of this species.

(a)   To find the mean of the sample use your calculator or the formula

$$\bar{x} = \frac{\sum xf}{\sum f} = \frac{626.25}{200} = 3.13125, \text{ so the sample mean is } 3.131\,25 \text{ kg}.$$

Since the sample mean is an unbiased estimator of the population mean, you can take this value as the estimate of the mean mass of all fish of this species.

Therefore the estimate of the mean mass of fish of this species is 3.13 kg, correct to 3 significant figures.

(b)   The variance of the sample is given by

$$\text{variance} = \frac{\sum x^2 f}{\sum f} - \bar{x}^2$$

$$= \frac{3220.1875}{200} - 3.131\,25^2 = 6.296\dots.$$

To obtain the unbiased estimate of the population variance, multiply the variance of the sample by $\dfrac{n}{n-1}$, which in this case is $\frac{200}{199}$.

$$s_{n-1}^2 = \frac{200}{199}\left(\frac{\sum x^2 f}{\sum f} - \bar{x}^2\right)$$

$$= \frac{200}{199} \times 6.296\dots = 6.33, \text{ correct to 3 significant figures.}$$

Exercise 10A

1   A random sample of 10 people working for a certain company with 4000 employees are asked, at the end of a day, how much they had spent on snacks that day. The results in £ are as follows.

   1.98   1.84   1.75   1.94   1.56   1.88   1.05   2.10   1.85   2.35

Calculate unbiased estimates of the mean and variance of the amounts spent on snacks that day by all workers employed by the company.

2   Fifty boxes of matches were selected at random from a large carton of such boxes. The number of matches in each box was counted. The results are summarised by $\sum x = 2400$, $\sum x^2 = 115\,212$. Calculate

   (a)   the mean and variance of the number of matches in a box for the sample of 50 boxes,

   (b)   unbiased estimates of the mean and variance of the number of matches in a box for all the boxes in the carton.

3   The diameters of 20 randomly chosen plastic doorknobs of a certain make were measured. The results, x cm, are summarised by $\sum(x-5) = 2.3$ and $\sum(x-5)^2 = 0.54$. Find

   (a)   the variance of the diameters in the sample,

   (b)   an unbiased estimate of the variance of the diameters of all knobs produced.

4   The number of vehicle accidents occurring along a long stretch of a particular motorway each day was monitored for a period of 100 randomly chosen days. The results are summarised in the following table.

Number of accidents	0	1	2	3	4	5	6
Number of days	8	12	27	35	13	4	1

Find unbiased estimates of the mean and variance of the daily number of accidents.

5   Unbiased estimates of the mean and variance of a population, based on a random sample of 24 observations, are 5.5 and 2.42 respectively. Another random observation of 8.0 is obtained. Find new unbiased estimates of the mean and variance with this new information.

6   A random sample of 150 pebbles was collected from a part of Brighton beach. The masses of the pebbles, correct to the nearest gram, are summarised in the following grouped frequency table.

Mass (g)	10–19	20–29	30–39	40–49	50–59	60–69	70–79	80–89
Frequency	1	4	22	40	49	28	4	2

Find, to 3 decimal places, unbiased estimates of the mean and variance of the masses of all the pebbles on this part of the beach.

10.4 The sampling distribution of the mean

In Chapter 5 you made use of the property that, when you take samples from a normal population with mean μ and known variance σ^2, the test statistic

$$Z = \frac{\overline{X} - \mu}{\frac{\sigma}{\sqrt{n}}}$$

has the standardised normal distribution $N(0,1)$. But what happens when you don't know the variance?

The obvious step is to replace σ^2 by the estimated variance S_{n-1}^2, and to use

$$T = \frac{\overline{X} - \mu}{\frac{S_{n-1}}{\sqrt{n}}}$$

as the test statistic.

But when you do this, the test statistic T no longer has a normal distribution. Because the estimate of the standard deviation varies from sample to sample, a second variable S_{n-1} has been introduced into the test statistic. This has the effect of slightly distorting the distribution. Instead of being normal, the test statistic has a distribution called **Student's t-distribution**, of just the **t-distribution**.

The reason for this name is that the distribution of T was studied by the British statistician W.S.Gosset (1876–1937) who wrote under the pen-name 'Student'. Much of Gosset's statistical work was done while he was employed at the Guinness brewery in Dublin.

You use the t-distribution in the same way as you use $N(0,1)$ when the variance is known. Calculators have programs for values of the probability density function and the cumulative distribution for the t-distribution, just as they do for the standardised normal distribution. Some also have programs for the inverse t-distribution. All you need to remember is:

> If the population variance is known, use the test statistic Z with the normal probability distribution.
>
> If the population variance is estimated from the variance of the sample, use the test statistic T with the t-probability distribution.

There is one complication. There is only one normal distribution with mean 0 and variance 1. But there are many different *t*-distributions, and which one you use depends on the size of the sample.

These *t*-distributions involve a positive integer parameter denoted by ν, which stands for the number of **degrees of freedom**. (The letter ν, pronounced 'nu', is the Greek letter 'n'.) When you use a calculator to find a probability with the *t*-distribution, you have to key in the value of ν as well as the number t. The value of t with ν degrees of freedom is sometimes denoted by t_ν.

The number of degrees of freedom depends on the particular way in which *t*-probability is being applied. When a *t*-test is used to test a hypothesis about the mean of a population, then $\nu = n - 1$, where n is the size of the sample.

> For a random variable from a normal population with mean μ the variable
>
> $T = \dfrac{\overline{X} - \mu}{\dfrac{S_{n-1}}{\sqrt{n}}}$ has a *t*-distribution with ν degrees of freedom, where $\nu = n - 1$.
>
> That is, $\dfrac{\overline{X} - \mu}{\dfrac{S_{n-1}}{\sqrt{n}}} \sim t_{n-1}$.

Fig. 10.2 shows the probability density graphs for the distributions t_3, t_5 and $N(0,1)$. For large values of ν the graph for t_ν is close to that of $N(0,1)$, but the graphs spread out more as ν decreases. This means that, in the 'tails', the probabilities are higher for t than for the normal distribution. This is what you should expect; the conclusion is more certain if it is based on a known value of the population variance than on an estimated one.

When you use a calculator the only difference between *t*-probability and standard normal probability is that you must also key in the value of ν. But to produce tables for *t*-probability like those of normal probability would require a substantial book, because you would need different tables for each ν.

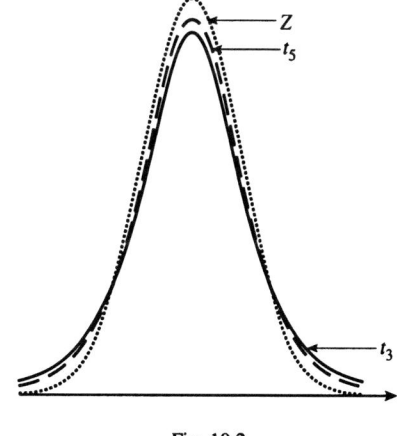

Fig. 10.2

To get round this, most *t*-probability tables are much less comprehensive, giving the values of t only at the most commonly used significance levels, with one row for each value of ν. You can carry out basic hypothesis testing with such tables, but you cannot use them to find *p*-values.

10.5 Hypothesis test of the population mean for a normal population

With a knowledge of the t-distribution it is now possible to extend hypothesis tests on the population mean to the situation in which a sample is taken from a normal population of unknown variance. In this situation the test statistic $\dfrac{\overline{X} - \mu}{\dfrac{S_{n-1}}{\sqrt{n}}}$ is distributed as t_{n-1} and the rejection region is found by using this

distribution. The following examples illustrate how rejection regions are found for one- and two-tail tests.

Example 10.5.1

In the past the mean lifetime, X (in hours), of a certain electrical component has been 10.4. A new manufacturing process is introduced which is designed to increase the lifetime. Experimental data collected from a random sample of components manufactured by the new process are summarised by $\sum x = 139.7$, $\sum x^2 = 1858.1$, $n = 11$. Making a suitable assumption, which you should state, test whether there is evidence at the 10% significance level that the new process has increased the lifetime.

This is a one-tail test looking for an increase. The null and alternative hypotheses are $H_0: \mu = 10.4$, $H_1: \mu > 10.4$.

The sample mean is $\overline{x} = \dfrac{\sum x}{n} = \dfrac{139.7}{11} = 12.7$,

and an unbiased estimate of the population variance is

$$S_{n-1}{}^2 = \frac{n}{n-1}\left(\frac{\sum x^2}{n} - \overline{x}^2\right)$$

$$= \frac{11}{10}\left(\frac{1858.1}{11} - 12.7^2\right) = 8.391\ldots\ .$$

The value of the test statistic is

$$t = \frac{\overline{x} - \mu}{\dfrac{S_{n-1}}{\sqrt{n}}} = \frac{12.7 - 10.4}{\dfrac{\sqrt{8.391\ldots}}{\sqrt{11}}} = 2.633\ldots\ .$$

There are now two ways to complete the test. Method 1 uses the inverse t-program, so if your calculator doesn't have this, use Method 2.

Method 1   For a one-tail test at the 10% significance level you want to find the value of t such that $P(T > t) = 0.10$. So $P(T \le t) = 0.90$, with $v = 11 - 1 = 10$, and the 'inverse t' program (or the t-distribution tables) gives the value $t = 1.372$. So the rejection region for this test is $T > 1.372$. This is illustrated in Fig. 10.3.

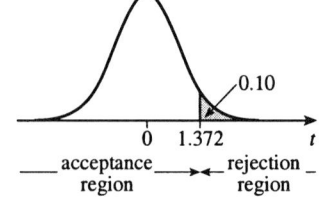

Fig. 10.3

The observed value 2.633 lies in this region, so the null hypothesis is rejected.

Method 2   Using the 't cumulative distribution' program, with $v = 10$,

$$P(T \le 2.633\ldots) = 0.9874\ldots\ ,$$

so $P(T > 2.633...) = 1 - 0.9874... = 0.0125...$, or about 1.3%.

Since this is less than 10%, the null hypothesis is rejected.

There is evidence at the 10% significance level that introducing the new process has increased the lifetime of the component.

Example 10.5.2
A student titrates 10 ml of 0.1 M acid against 0.1 M alkali five times and obtains the following results for the volume in ml of alkali.

9.88   10.18   10.23   10.39   10.25

Assuming that the volume of alkali used has a normal distribution, test at the 5% significance level whether these results show bias from the expected value of 10 ml.

Since the bias can be in either direction a two-tail test is appropriate. The null and alternative hypotheses are $H_0: \mu = 10$, $H_1: \mu \neq 10$.

The sample mean is $\bar{x} = \dfrac{\sum x}{n} = \dfrac{50.93}{5} = 10.186,$

and an unbiased estimate of the population variance is

$$s_{n-1}^{2} = \frac{n}{n-1}\left(\frac{\sum x^{2}}{n} - \bar{x}^{2}\right)$$

$$= \frac{5}{4}\left(\frac{518.9143}{5} - 10.186^{2}\right) = 0.035\,33.$$

The value of the test statistic is $t = \dfrac{\bar{x} - \mu}{\dfrac{s_{n-1}}{\sqrt{n}}} = \dfrac{10.186 - 10}{\dfrac{\sqrt{0.035\,33}}{\sqrt{5}}} = 2.2127...$.

A two-tail test at the 5% significance level has a rejection region divided into two parts, each corresponding to a probability of $2\frac{1}{2}\%$. The rejection and acceptance regions are shown in Fig. 10.4.

Method 1   The upper critical value is given by $P(T \leq t) = 0.975$ for 4 degrees of freedom. From the inverse *t* program, $t = 2.776$ and so the rejection region for this test is $|T| \geq 2.776$. The observed value of T does not lie in the rejection region and so the null hypothesis is not rejected.

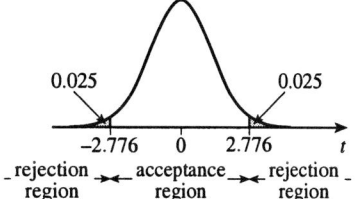

Fig. 10.4

Method 2   Using the *t* cumulative distribution with 4 degrees of freedom

$P(T \leq 2.2127...) = 0.954...$.

This value is less than 0.975, so T lies in the acceptance region.

The null hypothesis that the student's results do not show bias is accepted.

Example 10.5.3

The weights of packets of a certain brand of breakfast cereal are known to be distributed normally. It is claimed that the net weight of the contents is 450 g. The weights of the contents of seven packets are

445, 453, 447, 451, 440, 460, 449.

Is there any evidence at the 5% level that the population mean is less than 450 g?
What assumption must be made to carry out this test?

The sample mean is $\bar{x} = \dfrac{\sum x}{n} = \dfrac{3145}{7} = 449.28\ldots$, and an unbiased estimate of the population variance is

$$s_{n-1}^2 = \frac{n}{n-1}\left(\frac{\sum x^2}{n} - \bar{x}^2\right)$$

$$= \frac{7}{6}\left(\frac{1\,413\,245}{7} - 449.28\ldots^2\right) = 40.2381.$$

The null and alternative hypotheses are $H_0: \mu = 450$, $H_1: \mu < 450$.

The value of the test statistic is $t = \dfrac{\bar{x} - \mu}{\frac{s_{n-1}}{\sqrt{n}}} = \dfrac{449.28\ldots - 450}{\frac{\sqrt{40.2381}}{\sqrt{7}}} = -0.2979\ldots$, and the number of

degrees of freedom is $v = 7 - 1 = 6$.

Since the test statistic is negative, the p-value is $P(T \le -0.2979\ldots) = 0.3879\ldots$, or about 38.8%. This is much greater than 5%, so there is no evidence that the population mean is less than 450 g. The assumption made in carrying out this test is that the packets of cereal are randomly chosen.

Here is a summary of the method for carrying out a test of the population mean.

If a random sample of size n is drawn from a normal population of unknown variance, then the null hypothesis $H_0: \mu = \mu_0$ can be tested using the test statistic T. The value of T is given by

$$t = \frac{\bar{x} - \mu}{\frac{s_{n-1}}{\sqrt{n}}}, \text{ where } s_{n-1}^2 = \frac{n}{n-1}\left(\frac{\sum x^2}{n} - \bar{x}^2\right).$$

The rejection region for T depends on the form of the alternative hypothesis, H_1, the significance level and the number of degrees of freedom, $v = n - 1$.

- For a two-tail test at the $100\alpha\%$ significance level, the rejection region is given by $|T| \ge t$ where $P(T \le t) = 1 - \frac{1}{2}\alpha$.

- For a one-tail test for an increase at the $100\alpha\%$ significance level, the rejection region is given by $T \ge t$ where $P(T \le t) = 1 - \alpha$.

- For a one-tail test for a decrease at the $100\alpha\%$ significance level, the rejection region is given by $T \le -t$ where $P(T \le t) = 1 - \alpha$.

![Exercise 10B]

Exercise 10B

1   A new method for determining the pH value of soil has been developed and is used on a random sample of 15 specimens taken from an area whose pH value has a mean of 6.32. The 15 values give a sample mean of $\bar{x} = 6.14$ and an estimated population variance of $s_{n-1}^2 = 10$. Assuming that the pH values in the area have a normal distribution, test, at the 5% significance level, whether the mean pH value as determined by the new method differs from 6.32.

2   The time taken for a machine to produce a certain plastic bowl has a normal distribution with mean 4.8 seconds. An adjustment is made to the machine which is intended to reduce the mean, and in order to test that it is successful, the times taken to produce 8 bowls are measured. The results, t seconds, are as follows:

   4.2   4.5   4.9   4.3   4.5   4.7   4.6   4.4.

   Test, at the $2\frac{1}{2}\%$ significance level, whether or not the adjustment had the desired effect.

3   The plate glass used in a construction project is required to have a mean thickness greater than 5 mm in order to be acceptable. The thickness of each of 25 samples of a certain kind of plate glass is measured, giving a sample mean of 5.14 mm and estimated population variance of 0.29^2 mm^2. Assuming that the thickness of the glass is distributed normally, carry out a test, at the 5% significance level, to decide whether or not the sampled kind of plate glass is acceptable.

4   The number of minutes late that a Number 38 bus arrives at a city centre stop has a mean of 5.7 during the 'rush hour'. After a reorganisation of the traffic lights in the city, the times of arrival of six Number 38 buses at the city centre bus stop were noted. The number of minutes late were as follows, a negative value indicating that the bus was early.

   2.3   5.6   −1.8   4.3   −1.1   3.2

   Assuming that these times comprise a random sample selected from a normal distribution, test, at the 1% significance level, whether or not the new mean is less than 5.7.

5   Haemoglobin levels in females may be modelled by a normal distribution with mean 14.2 (grams per decilitre). As part of a health study, 10 randomly chosen female students from a college had their haemoglobin levels, h, measured. Results summaries are $\sum h = 147.9$ and $\sum h^2 = 2203.19$. Test, at the 5% significance level, whether the mean haemoglobin level of female students in the college differs from the mean level of all females.

6   The mean birth weights of 20 babies born during April 2001 in a London hospital maternity unit was 3.207 kg and gave an unbiased estimate of the population variance of $s_{n-1}^2 = 0.461^2$ kg^2. It may be assumed that these babies form a random sample of all those born in London hospitals during 2001, whose birth weights are distributed normally with mean μ kg. A test of the null hypothesis $\mu = \mu_0$ against the alternative hypothesis $\mu \neq \mu_0$ is carried out at the 10% significance level. Find the set of values of μ_0 for which the null hypothesis is accepted.

10.6   Application to matched pairs

Suppose you want to see whether the time taken to complete a simple task decreases with practice. One way of investigating this would be to take a random sample of people and measure the times which each person takes to perform the task on their first and second attempts. An example of such data is shown in Table 10.5.

Person	A	B	C	D	E	F	G	H
First attempt	6.3	3.5	7.1	3.7	8.4	3.9	4.7	5.2
Second attempt	5.1	3.4	6.2	4.5	7.3	4.0	3.6	5.1
Difference, d	1.2	0.1	0.9	−0.8	1.1	−0.1	1.1	0.1

Table 10.5

The null hypothesis is $H_0: \mu_x = \mu_y$, where μ_x is the mean time taken by all people on their first attempt and μ_y is the mean time taken by all people on their second attempt. This can be written as $H_0: \mu_x - \mu_y = 0$. The data collected takes the form of pairs of values, one pair for each person. You can see that there is quite a lot of variation between different people. For example, B is fast on both attempts whereas E is slow. However, it is not the variation between individuals which is of interest but their improvement with practice. If practice does decrease the time taken, then you would expect the first value to be greater than the second. This means that the improvement can be measured by the difference, d, between the first and the second values. The values of these differences are given in the last line of the table. You can check that the average difference, $\bar{d}$, is 0.45 minutes.

The question which now needs to be answered is whether this value indicates that the second time is significantly lower than the first. To do this you need to consider the distribution of $D = X - Y$, where X is the time on the first attempt and Y is the time on the second attempt.

The mean, μ_d, of the sampling distribution of D is given by

$$\mu_d = E(D) = E(X - Y)$$
$$= E(X) - E(Y)$$
$$= \mu_x - \mu_y.$$

The null hypothesis $H_0: \mu_x - \mu_y = 0$ is thus equivalent to $H_0: \mu_d = 0$. This is just the form of the null hypothesis for a single sample which you met in Section 10.5. So, provided that D has a normal distribution, this null hypothesis can be tested using the statistic T with $n - 1$ degrees of freedom. The variance of the sampling distribution of D is not known, but it can be estimated by S_d^2, where S_d^2 is the unbiased estimator. For the data in Table 10.4 the unbiased estimate of the variance of the differences is $s_d^2 = 0.531\ldots$. You can then calculate

$$t = \frac{\bar{d} - 0}{\frac{s_d}{\sqrt{n}}} = \frac{0.45 - 0}{\frac{\sqrt{0.531\ldots}}{\sqrt{8}}} = 1.7459\ldots \; .$$

The alternative hypothesis is $H_1: \mu_x > \mu_y$, or $\mu_x - \mu_y > 0$. You therefore have to do a one-tail test with $v = 7$. You might choose to use a 10% significance level.

To complete the test you can find, using a calculator or tables, that the rejection region is $T > 1.415$ and that the observed value $t = 1.7459\ldots$ lies in this region; or you can calculate $P(T > 1.7459\ldots) = 0.062\,16\ldots$, and note that 6.22% is less than 10%. The null hypothesis is therefore rejected, and the alternative hypothesis that people take less time with practice is accepted.

A test like this is called a **matched pairs test.** *The word 'matched' is used to indicate that the same people participate in the second trial as in the first.*

Example 10.6.1
To investigate the difference in wear on front and rear tyres of motorcycles, 50 motorcycles of the same model were fitted with new tyres of the same brand. After the motorcycles had been driven for 2000 miles the depths of tread on the front and rear tyres were measured in mm. For each motorcycle the value of d = (depth of front tread – depth of rear tread) was calculated. The results can be summarised by $\sum d = 4.7$ and $\sum d^2 = 3.98$. Test, at the 5% significance level, whether there is a difference in wear on the front and rear tyres.

This is an example of an experiment where the data are collected in pairs. The null hypothesis is $H_0 : \mu_x = \mu_y$ where μ_x is the population mean for front tyres and μ_y the population mean for rear tyres. The alternative hypothesis is $H_1 : \mu_x \neq \mu_y$.

$$\bar{d} = \frac{4.7}{50} = 0.094 \quad \text{and} \quad s_{n-1}{}^2 = \frac{50}{49}\left(\frac{3.98}{50} - 0.094^2\right) = 0.0722\ldots\ .$$

This gives the value of the test statistic $t = \dfrac{0.094 - 0}{\dfrac{\sqrt{0.0722\ldots}}{\sqrt{50}}} = 2.47\ldots\ .$

Method 1   To find the two-tail rejection region you want the number c such that $P(|T| > c) = 0.05$, so that $P(T > c) = 0.025$. With $\nu = 49$ the inverse *t*-program gives $c = 2.010$. (If you are using tables you may have to get this value by interpolation between 2.021 for $\nu = 40$ and 2.000 for $\nu = 60$.) So the rejection region is $|T| > 2.010$. The calculated value $t = 2.47\ldots$ lies in this region.

Method 2   The cumulative *t*-program gives $P(T > 2.47\ldots) = 0.008\,44\ldots$, so that $P(|T| > 2.47\ldots) = 2 \times 0.008\,44\ldots = 0.016\,89\ldots$, or 1.7%. This is less than the 5% significance level.

Thus H_0 is rejected and H_1, that there is different wear on the front and rear tyres, is accepted.

A matched pairs test can also be used to test whether the mean of paired differences is some constant value, expressed by the null hypothesis $H_0 : \mu_x - \mu_y = c$. In this case the test statistic is $T = \dfrac{\bar{D} - c}{\dfrac{S_d}{\sqrt{n}}}$.

Exercise 10C

1   Some psychologists believe that the IQ of the first-born child in a family is significantly greater than the IQ of the last born. In order to investigate this belief, a random sample of 8 families with more than one child agreed to allow their children's IQs to be measured, with the following results.

Family	A	B	C	D	E	F	G	H
IQ of first born	97	121	89	112	138	125	104	114
IQ of last born	101	116	97	108	130	121	101	105

Assuming that the differences have a normal distribution test the psychologists' belief using a 5% significance level.

2 A person's systolic blood pressure is a measure of the pressure exerted by the heart when it contracts and pushes blood around the body. When the heart has just ceased to contract and is dilating ready for the next contraction, the blood pressure drops and is called the diastolic pressure.

The following table gives the systolic and diastolic blood pressures (measured in mm of mercury) of 6 randomly chosen people with diabetes.

Patient	A	B	C	D	E	F
Systolic pressure	141	129	117	115	93	101
Diastolic pressure	83	76	71	59	51	64

Let D denote the amount by which the systolic pressure exceeds the diastolic pressure of a randomly chosen person with diabetes, and let μ_D denote the mean of D. Assuming that D has a normal distribution, test the hypothesis $\mu_D > 40$ at the 5% significance level.

3 The reaction times taken by 10 motorists to apply the brakes of their cars were measured when the motorists had not drunk any alcohol, and after they had drunk a measured amount of alcohol. The reaction times, in hundredths of a second, are given in the following table.

Motorist	A	B	C	D	E	F	G	H	I	J
Without alcohol	40	25	19	23	38	37	28	37	41	27
With alcohol	50	37	35	34	52	50	40	46	53	38

Assuming that the population of differences has a normal distribution, test, at the 1% significance level, whether the drinking of alcohol increases the mean reaction time by more than 0.1 s.

4 An experiment was carried out to compare the difference in the effects of organic and chemical fertilisers on potato yields. Eleven plots of land were selected and two seed potatoes were grown on each plot at a distance of 10 m apart. On one potato an organic fertiliser was used, and on the other, a chemical fertiliser. The choice of which to use was decided by tossing a coin. The differences in yields, d grams, where d = (mass of organic crop – mass of chemical crop), are summarised by

$$\sum d = -310 \quad \text{and} \quad \sum d^2 = 208\,702.$$

Assuming that the differences have a normal distribution, test, at the 5% significance level, whether there is a difference between the population mean yields.

5 A study of the effect of vitamins on attention span was carried out on 40 sets of identical twins of the same age. One twin was randomly chosen to have the vitamin pill and the other was given a placebo (a pill with no vitamin). Each twin was given a puzzle to solve and the time, in minutes, that each twin remained with the puzzle was measured. The difference in time, d minutes, where d = (vitamin time – placebo time), has mean $\bar{d} = 2.92$ and the unbiased estimate of the variance is 141.23.

If it could be assumed that the differences are distributed normally, what would be the conclusion of the test?

11 Confidence intervals

This chapter introduces the idea of a confidence interval. When you have completed it, you should be able to

- determine a confidence interval for the population mean in the context of a sample drawn from a normal population of known variance
- determine a confidence interval for the population mean in the context of a large sample drawn from any population of known variance
- use a t-distribution, in the context of a sample drawn from a normal population, to determine a confidence interval of the population mean when the variance is unknown
- determine, from a large sample, an approximate confidence interval for a population proportion.

11.1 The concept of a confidence interval

It was shown in Section 10.1 that the mean, $\overline{X}$, of a random sample is an unbiased estimator of the population mean, μ. For example, to estimate the mean amount of pocket money received by all the children in a primary school you could take a random sample of the children and ask each child how much pocket money he or she receives each week. Suppose you find that $\sum x = 111.50$ for a random sample of 50 children, where x is measured in £. Then an unbiased estimate of the population mean, μ, is given by

$$\overline{x} = \frac{\sum x}{n} = \frac{111.50}{50} = 2.23.$$

Such a value is called a **point estimate** because it gives an estimate of the population mean in the form of a single value or 'point' on a number line. Such values are useful, for example, in comparing populations. However, since $\overline{X}$ is a random variable, the value which it takes will vary from sample to sample. As a result you have no idea how close to the actual population mean a point estimate is likely to be. The purpose of a 'confidence interval' is to give an estimate in a form which also indicates the estimate's likely accuracy. A **confidence interval of the mean** is a range of values which has a given probability of 'trapping' the population mean. It is usually taken to be symmetrical about the sample mean. So, if the sample mean takes the value $\overline{x}$, the associated confidence interval would be $[\overline{x} - c, \overline{x} + c]$ where c is a number whose value has yet to be found.

Notice the notation for an interval used here. The interval $[\overline{x} - c, \overline{x} + c]$ means the real numbers from $\overline{x} - c$ to $\overline{x} + c$, including the end-points. For example, $[-2.1, 6.8]$ means real numbers y such that $-2.1 \le y \le 6.8$. See Higher Level Book 1 Section 22.4.

In Fig. 11.1 the sample mean $\overline{X}$ takes a value $\overline{x}_1$ and the confidence interval covers the range of values $[\overline{x}_1 - c, \overline{x}_1 + c]$.

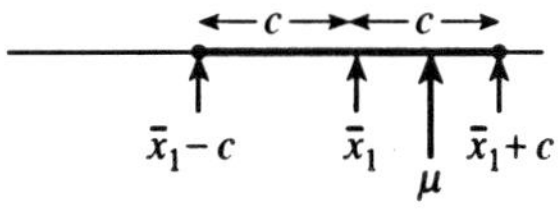

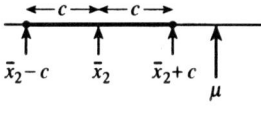

Fig. 11.1

Fig. 11.2

In this case the confidence interval traps the population mean, μ. For a different sample, with a different sample mean, $\overline{x}_2$, the confidence interval might not trap μ. This situation is illustrated in Fig. 11.2.

The end-points of the confidence interval are themselves random variables since they vary from sample to sample. They can be written as $\overline{X} - c$ and $\overline{X} + c$. You can see from Fig. 11.1 and Fig. 11.2 that the confidence interval will trap μ if the difference between the sample mean and the population mean is less than or equal to c. Expressed algebraically this condition is $\left| \overline{X} - \mu \right| \leq c$.

The next section explains how the value of c is chosen so as to give a specified probability that the confidence interval traps μ.

11.2 Calculating a confidence interval

Consider the following situation. The masses of tablets produced by a machine are known to be distributed normally with a standard deviation of 0.012 g. The mean mass of the tablets produced is monitored at regular intervals by taking a sample of 25 tablets and calculating the sample mean, $\overline{X}$.

Suppose you wish to find an interval which has a 95% probability of trapping the population mean, μ. The mass, X (in grams), of a single tablet is distributed normally with unknown mean μ and standard deviation 0.012; that is, $X \sim N\left(\mu, 0.012^2\right)$. So, for a sample of size 25, $\overline{X} \sim N\left(\mu, \dfrac{0.012^2}{25}\right)$.

Using the notation of the previous section, the population mean is trapped in the interval $\left[\overline{X} - c, \overline{X} + c\right]$, where c is a constant, if $\left| \overline{X} - \mu \right| \leq c$. The interval has a probability of 95% of trapping the population mean if, and only if, $P\left(\left| \overline{X} - \mu \right| \leq c\right) = 0.95$.

Fig. 11.3 shows the sampling distribution of $\overline{X}$ with this probability indicated. The value of c is found by standardising $\left| \overline{X} - \mu \right|$ to give $Z = \dfrac{\overline{X} - \mu}{\dfrac{0.012}{\sqrt{25}}}$ where $Z \sim N(0,1)$.

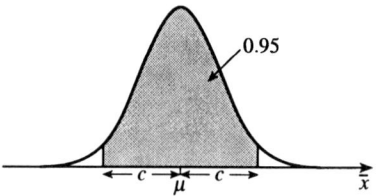

Fig. 11.3

Fig. 11.4 shows the distribution of Z with the probability of 0.95 indicated. If your calculator can operate in two-tail mode, use this to find z such that $P\left(\left| Z \right| \leq z\right) = 0.95$; if not, convert this to a cumulative normal distribution condition, and find z such that $P(Z < z) = 0.975$. Either way, $z = 1.960$.

Thus $1.960 = \dfrac{\overline{X} - \mu}{\dfrac{0.012}{\sqrt{25}}}$.

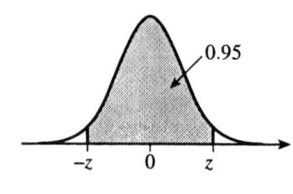

Fig. 11.4

But $\overline{X} - \mu = c$, so

$$1.960 = \frac{c}{\dfrac{0.012}{\sqrt{25}}}$$

giving

$$c = 1.960 \times \frac{0.012}{\sqrt{25}} = 0.004\ 70, \text{ correct to 3 significant figures.}$$

So the interval which has a 95% probability of trapping μ is $\left[\overline{X} - 0.0047, \overline{X} + 0.0047\right]$.

Suppose that you took a sample of 25 tablets and found that the mean mass was, for example, 0.5642 g. This sample mean would give a value for the interval of $[0.5642 - 0.0047, 0.5642 + 0.0047]$, which is $[0.5595, 0.5689]$. Such an interval is called a **95% confidence interval of the population mean**.

It is important to realise that such a confidence interval may or may not trap μ, depending on the value of $\overline{X}$. Suppose, for a moment, that you know the value of μ and you take a number of different samples of 25 tablets. Fig. 11.5 shows confidence intervals calculated in the way described above for 30 different samples of 25 tablets.

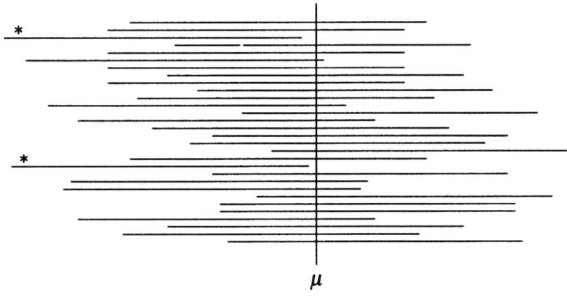

μ

Fig. 11.5

Most of the confidence intervals trap μ but a few (marked *) do not. On average, the proportion of 95% confidence intervals which trap μ is 95%. In practice, of course, you do not know μ and would usually take only one sample and from it calculate one confidence interval. In this situation, you cannot *know* whether your particular confidence interval does include μ: all you can say is that, on average, 95 times out of 100 it will contain μ.

The method which has been described for finding a 95% confidence interval of the mean can be generalised to a sample of size n from a normal population with standard deviation σ. Replacing 25 by n and 0.012 by σ gives

$$1.960 = \frac{c}{\frac{\sigma}{\sqrt{n}}},$$

which, on rearranging, gives $c = 1.960 \dfrac{\sigma}{\sqrt{n}}$.

> Given a sample of size n from a normal population with variance σ^2, a 95% confidence interval for the population mean is given by
>
> $$\left[\overline{x} - 1.960\frac{\sigma}{\sqrt{n}}, \overline{x} + 1.960\frac{\sigma}{\sqrt{n}}\right],$$
>
> where $\overline{x}$ is the sample mean.

Example 11.2.1

The lengths of nails produced by a machine are known to be distributed normally with mean μ mm and standard deviation 0.7 mm. The lengths, in mm, of a random sample of 5 nails are 107.29, 106.56, 105.94, 106.99, 106.47.

(a) Calculate a symmetric 95% confidence interval for μ, giving the end-points to 1 decimal place.

(b) Two hundred random samples of 5 nails are taken and a symmetric 95% confidence interval for μ is calculated for each sample. Find the expected number of intervals which do not contain μ.

(a)   The mean of the sample is given by

$$\bar{x} = \frac{107.29 + 106.56 + 105.94 + 106.99 + 106.47}{5} = 106.65.$$

Substituting this value of $\bar{x}$ together with $\sigma = 0.7$ and $n = 5$ into the expression in the shaded box gives a symmetric 95% confidence interval for μ of

$$\left[106.65 - 1.960 \times \frac{0.7}{\sqrt{5}}, \ 106.65 + 1.960 \times \frac{0.7}{\sqrt{5}} \right] = [106.65 - 0.61, 106.65 + 0.61].$$

The symmetric 95% confidence interval for μ, measured in mm, is $[106.0, 107.3]$.

(b)   On average, 95% of the confidence intervals should include μ. This means that 5% will not include μ. So out of 200 confidence intervals you would expect $200 \times 5\% = 10$ not to include μ.

It is not easy to suggest a general rule about the accuracy to which confidence intervals are given, so you have to exercise common sense. In Example 11.2.1, since the standard deviation is given to only 1 significant figure, you could not justify giving the bounds of the confidence interval to more than 1 decimal place. But if the standard deviation were known more precisely, an answer correct to 2 decimal places would be appropriate.

Example 11.2.2

For a method of measuring the velocity of sound in air, the results of repeated experiments are known to be distributed normally with standard deviation 6 m s^{-1}. A number of measurements are made using this method, and from these measurements a symmetric 95% confidence interval for the velocity of sound in air is calculated. Find the width of this confidence interval for   (a) 4,   (b) 36 measurements.

A symmetric 95% confidence interval extends from $\bar{x} - 1.960 \dfrac{\sigma}{\sqrt{n}}$ to $\bar{x} + 1.960 \dfrac{\sigma}{\sqrt{n}}$, so its width is

$$\bar{x} + 1.960 \frac{\sigma}{\sqrt{n}} - \left(\bar{x} - 1.960 \frac{\sigma}{\sqrt{n}} \right) = 2 \times 1.960 \frac{\sigma}{\sqrt{n}}.$$

In this example $\sigma = 6$, so the width of the confidence interval is $2 \times 1.960 \dfrac{6}{\sqrt{n}} = \dfrac{23.52}{\sqrt{n}}$.

(a)   For $n = 4$, the width of the confidence interval is $\dfrac{23.52}{\sqrt{4}} = 11.76.$

(b)   For $n = 36$, the width of the confidence interval is $\dfrac{23.52}{\sqrt{36}} = 3.92.$

Nine times as many measurements are needed to reduce the confidence interval width by a factor of three.

11.3 Different levels of confidence

As Example 11.2.1 makes clear, there is a probability of 5 in 100 that a 95% confidence interval does not include μ. There are circumstances in which you may wish to be more certain that the confidence interval which you have calculated does include μ. For example, you may wish to be 99% certain. Fig. 11.6 is the diagram corresponding to Fig. 11.3 for this situation. The only difference in the calculation of the confidence interval is that a different value of z is needed. In this case $P(|Z| \le z) = 0.99$ or $P(Z \le z) = 0.995$, giving $z = 2.576$, so that the 99% confidence interval of the mean is given by

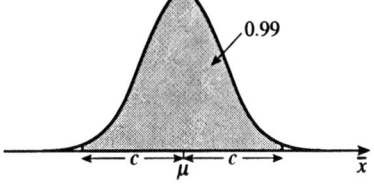

Fig. 11.6

$$\left[\bar{x} - 2.576 \frac{\sigma}{\sqrt{n}}, \bar{x} + 2.576 \frac{\sigma}{\sqrt{n}} \right].$$

Note that this 99% confidence interval is wider than the 95% confidence interval: this is to be expected since the former is more likely to trap μ than the latter. You can see that there is a balance between precision and certainty: if you increase one you decrease the other. This is similar to the balance between Type I and Type II errors in hypothesis testing (see Section 9.2). The only way of decreasing the probabilities of both types of error simultaneously is to increase the sample size. In the same way, to increase both the precision and the certainty of a confidence interval you have to increase the sample size.

For a 90% confidence interval, the appropriate value of z is that value for which $P(|Z| \le z) = 0.90$ or $P(Z \le z) = 0.95$, giving $z = 1.645$. You may have spotted that the values of z used in confidence intervals correspond to those used in hypothesis testing. To generalise, the critical value of z for a two-tail test at the $100\alpha\%$ significance level is the same as the value of z used to calculate a symmetric $100(1 - \alpha)\%$ confidence interval (see Fig. 11.7). The critical value of z for a two-tail test at the $100\alpha\%$ level is also the same as the critical value of z for a one-tail test at the $\frac{1}{2} \times 100\alpha\%$ level. These values are given in the shaded box at the end of this section where the probabilities refer to the acceptance region.

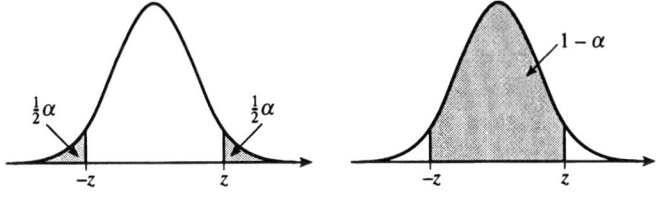

Fig. 11.7

A $100(1 - \alpha)\%$ confidence interval of the population mean for a sample of size n taken from a normal population with variance σ^2 is given by

$$\left[\bar{x} - z \frac{\sigma}{\sqrt{n}}, \bar{x} + z \frac{\sigma}{\sqrt{n}} \right],$$

where $\bar{x}$ is the sample mean and the value of z is such that $P(|Z| \le z) = 1 - \alpha$ or $P(Z \le z) = 1 - \frac{1}{2}\alpha$.

Example 11.3.1

The masses of sweets produced by a machine are normally distributed with a standard deviation of 0.5 grams. A sample of 50 sweets had a mean mass of 15.21 grams.

(a) Find a 99% confidence interval for μ, the mean mass of all sweets produced by the machine.

The manufacturer of the machine claims that it produces sweets with a mean mass of 15 grams.

(b) State, giving a reason, whether the confidence interval calculated in part (a) supports this claim.

(a)   From the results given in the shaded box above, a $100(1-\alpha)\%$ confidence interval for μ is given by $\left[\bar{x} - z\dfrac{\sigma}{\sqrt{n}}, \bar{x} + z\dfrac{\sigma}{\sqrt{n}} \right]$ where the value of z is such that $P(Z \le z) = 1 - \tfrac{1}{2}\alpha$.

For a 99% confidence interval, $100(1-\alpha) = 99$, giving $\alpha = 0.01$. Thus the value of z required is that for which

$$P(Z \le z) = 1 - \tfrac{1}{2}\alpha$$
$$= 1 - \tfrac{1}{2} \times 0.01 = 0.995.$$

From the inverse normal cumulative distribution program, $z = 2.576$.

Substituting $z = 2.576$, $\bar{x} = 15.21$, $\sigma = 0.5$ and $n = 50$ into the expression for the confidence interval gives a 99% confidence interval of

$$\left[15.21 - 2.576 \times \frac{0.5}{\sqrt{50}}, 15.21 + 2.576 \times \frac{0.5}{\sqrt{50}} \right] = [15.027\ldots, 15.392\ldots]$$
$$= [15.03, 15.39], \text{ correct to 2 decimal places.}$$

(b)   The confidence interval does not include the value 15, which suggests that the manufacturer's claim is untrue.

You can check that the same conclusion is reached if a hypothesis test is carried out at the 1% significance level with $H_0: \mu = 15$ and $H_1: \mu \ne 15$.

Example 11.3.2

The measurement error made in measuring the concentration in parts per million (ppm) of nitrate ions in water by a particular method is known to be distributed normally with mean 0 and standard deviation 0.05.

(a) If 10 measurements on a specimen gave $\sum x = 11.37$ ppm, determine a symmetric 99.5% confidence interval for the true concentration, μ, of nitrate ions in the specimen.

(b) How many measurements would be required in order to reduce the width of this interval to 0.03 ppm at most?

(a)   The measured value, X, of the nitrate ion concentration is equal to $\mu + Y$ where Y is the measurement error. Thus

$$E(X) = E(\mu + Y)$$
$$= \mu + E(Y)$$
$$= \mu + 0 = \mu,$$

and

$$Var(X) = Var(\mu + Y)$$
$$= Var(\mu) + Var(Y)$$
$$= 0 + 0.05^2 = 0.05^2.$$

This means that $X \sim N(\mu, 0.05^2)$.

For a 99.5% confidence interval, $1 - \alpha = 0.995$, so $\alpha = 0.005$ and $\frac{1}{2}\alpha = 0.0025$.

Thus $P(|Z| \le z) = 1 - \alpha = 0.995$, or

$$P(Z \le z) = 1 - \frac{1}{2}\alpha$$
$$= 1 - 0.0025 = 0.9975$$

which gives

$$z = 2.807.$$

So a 99.5% confidence interval for μ is

$$\left[\bar{x} - 2.807\,\frac{\sigma}{\sqrt{n}}, \bar{x} + 2.807\,\frac{\sigma}{\sqrt{n}}\right] = \left[\frac{11.37}{10} - 2.807 \times \frac{0.05}{\sqrt{10}}, \frac{11.37}{10} + 2.807 \times \frac{0.05}{\sqrt{10}}\right]$$
$$= [1.137 - 0.044,\ 1.137 + 0.044]$$
$$= [1.093, 1.181].$$

(b)   The width of the 99.5% confidence interval for a sample of size n is

$$2 \times 2.807 \times \frac{0.05}{\sqrt{n}}.$$

For this width to be at most 0.03, $2 \times 2.807 \times \frac{0.05}{\sqrt{n}} \le 0.03$.

Rearranging gives

$$\sqrt{n} \ge \frac{2 \times 2.807 \times 0.05}{0.03}, \quad \text{so} \quad n \ge 87.5\ldots\ .$$

Since n must be an integer, 88 or more measurements are required to give a confidence interval of width 0.03 ppm at most.

Some values of z for commonly used confidence intervals are given below:

Confidence interval	z
90%	1.645
95%	1.960
98%	2.326
99%	2.576

11.4 Confidence intervals for non-normal populations

So far all the examples of confidence intervals have been obtained from samples drawn from populations with a normal distribution. But if the sample is large enough, that condition can be relaxed.

The reasoning used to establish confidence intervals in Section 11.2 was based on the fact that in a normal population the sample mean $\overline{X}$ is distributed normally. But you know from the central limit theorem (Section 7.1) that, if the sample is large enough, the sample mean has a distribution that is approximately normal whatever the distribution of the underlying population. This means that you can use the same method to find a confidence interval for the mean of any population, provided that the sample is large enough for the normal approximation to be valid.

Example 11.4.1

A supermarket chain has a flexible hours policy for its regular check-out staff. The management is not willing to publish the mean hours worked, but has made it known that the effect of the policy is that the standard deviation of the hours worked per week is 2.9 hours. A trade union researcher asks a random sample of 80 of these employees to record how many hours they work in a typical week. She finds that the total of the working times recorded is 3064 hours. Use these data to find a 90% confidence interval for the mean number of hours per week worked by members of the check-out staff.

The distribution of hours worked by different members of staff is not known, but the sample size of 80 is large enough for the researcher to assume that the distribution of the sample mean is approximately normal.

The mean number of hours worked by a single member of staff is $3064 \div 80$, which is 38.30. The shaded box at the end of Section 11.3 gives the value of z for a 90% confidence interval to be 1.645. The confidence interval is therefore given by

$$\left[38.30 - 1.645 \times \frac{2.9}{\sqrt{80}}, 38.30 + 1.645 \times \frac{2.9}{\sqrt{80}} \right],$$

which is $[37.77, 38.83]$, correct to 2 decimal places.

Exercise 11A

1   Bags of sugar have masses which are distributed normally with mean μ grams and standard deviation 4.6 grams. The sugar in each of a random sample of 5 bags taken from a production line is weighed, with the following results, in grams.

    498.2   501.3   503.7   496.8   502.5

Calculate a symmetric 95% confidence interval for μ.

If a 95% symmetric confidence interval for μ was calculated for each of the 200 samples of 5 bags, how many of the confidence intervals would be expected to contain μ?

2   The volume of milk in litre cartons filled by a machine has a normal distribution with mean μ litres and standard deviation 0.05 litres. A random sample of 25 cartons was selected and the contents, x litres, measured. The results are summarised by $\sum x = 25.11$. Calculate

    (a)   a symmetric 98% confidence interval for μ,

    (b)   the width of a symmetric 90% confidence interval for μ based on the volume of milk in a random sample of 50 cartons.

3 The random variable X has a normal distribution with mean μ and variance σ^2. A symmetric 90% confidence interval for μ based on a random sample of 16 observations of X has width 4.24. Find

(a) the value of σ,

(b) the width of a symmetric 90% confidence interval for μ based on a random sample of 4 observations of X,

(c) the width of a symmetric 95% confidence interval for μ based on a random sample of 4 observations of X.

4 The heights of fully-grown British males may be modelled by a normal distribution with mean 178 cm and standard deviation 7.5 cm. The 11 male (fully-grown) biology students present at a university seminar had a mean height of 175.2 cm. Assuming a standard deviation of 7.5 cm, and stating any further assumption, calculate a symmetric 99% confidence interval for the mean height of all fully-grown male biology students.

Does the confidence interval suggest that fully-grown male biology students have a different mean height from 178 cm?

5 A machine is designed to produce metal rods of length 5 cm. In fact, the lengths are distributed normally with mean 5.00 cm and standard deviation 0.032 cm. The machine is moved to a new site and, in order to check whether or not the mean length has altered, the lengths of a random sample of 8 rods are measured. The results, in cm, are as follows.

   5.07 4.95 4.98 5.06 5.13 5.05 4.98 5.06

(a) Assuming that the standard deviation is unchanged, calculate a symmetric 95% confidence interval for the mean length of the rods produced by the machine in its new position.

(b) State, giving a reason, whether you consider that the mean length has changed.

6 A method used to determine the percentage of nitrogen in a fertiliser has an error which is distributed normally with zero mean and standard deviation 0.34%. Ten independent determinations of the percentage of nitrogen gave a mean value of 15.92%.

(a) Calculate a symmetric 98% confidence interval for the percentage of nitrogen in the fertiliser.

(b) Find the smallest number of extra independent determinations that would reduce the width of the symmetric 98% confidence interval to at most 0.4%.

7 Each year a city sets a general knowledge test to all its 10-year-old children. The standard deviation of the marks is always about 23, and the aim is to have a mean mark of 70. This year's test is tried out on 200 randomly chosen children in a neighbouring city, and the mean mark achieved in the trial is 67.8. Find a 90% confidence interval for the true mean for this test. What action should be taken as a result of the trial?

11.5 Populations with unknown variance

So far it has been assumed that the variance of the underlying population is known. In practice you will often not know this variance, and you will have to estimate it from the variance of the sample.

This is just the situation you met in Section 10.5 when testing hypotheses about the population mean, and you deal with it in the same way. That is, the population variance σ^2 is replaced by the estimated population variance s_{n-1}^2, and the multiplier is found from the t-distribution rather than the normal distribution. The rule is then:

> Given a sample of size n from a normal population of unknown variance, a $100(1-\alpha)\%$ confidence interval for the population mean is given by
>
> $$\left[\bar{x} - t\frac{s_{n-1}}{\sqrt{n}}, \bar{x} + t\frac{s_{n-1}}{\sqrt{n}} \right],$$
>
> where $\bar{x}$ is the sample mean and the value of t is such that $P(|T| \leq t) = 1 - \alpha$, or $P(T \leq t) = 1 - \frac{1}{2}\alpha$, for $\nu = n - 1$ degrees of freedom.

Example 11.5.1

Ten measurements of the zero error on an ammeter yielded the results $+0.13$, -0.09, $+0.06$, $+0.15$, -0.02, $+0.03$, $+0.01$, -0.02, -0.07, $+0.05$. (The zero error is the reading when no electric current is passing through the ammeter.) Assuming that these measurements come from a normal population, calculate a 95% confidence interval for the mean zero error.

First it is necessary to calculate the sample mean and an unbiased estimate of the population variance from the given measurements.

$$\bar{x} = \frac{\sum x}{n} = \frac{0.23}{10} = 0.023,$$

and

$$s_{n-1}^2 = \frac{n}{n-1}\left(\frac{\sum x^2}{n} - \bar{x}^2 \right) \text{ so } s_{n-1} = 0.078\,18\ldots .$$

For a 95% confidence interval, $\alpha = 0.05$, so $P(T \leq t) = 1 - \frac{1}{2} \times 0.05 = 0.975$ for $\nu = n - 1 = 9$ degrees of freedom. Using the inverse cumulative t-distribution program, the value of t is 2.262; so a 95% confidence interval for the population mean is

$$\left[0.023 - 2.262 \times \frac{0.078\,18\ldots}{\sqrt{10}}, 0.023 + 2.262 \times \frac{0.078\,18\ldots}{\sqrt{10}} \right] = [-0.0329, 0.0789],$$

where the answers have been given to 3 significant figures.

You can also find the value of t for a given confidence level from a table of critical values. In this example you will find the value $t = 2.262$ tabulated for $P(X \leq t) = 0.975$ with $\nu = 9$.

Exercise 11B

1   A symmetric $c\%$ confidence interval for a population mean is to be calculated using a random sample of n observations of the normal random variable X, which has unknown mean and unknown variance. State the values of t used in the following cases.

(a) $c = 90$, $n = 12$ (b) $c = 95$, $n = 10$ (c) $c = 98$, $n = 25$ (d) $c = 99.5$, $n = 100$

2   Bottles of El Bombero 1999, a Spanish red wine, were advertised as having 'an incredible 15%' alcohol content. Wine from a random sample of six bottles was analysed and gave percentage alcohol contents of   14.6, 15.1, 14.7, 15.3, 14.9, 15.0.

Stating any required assumption, calculate a symmetric 95% confidence interval for the mean percentage alcohol content in all bottles of El Bombero 1999.

3   The number of calls made in May 2000 to each of 20 randomly chosen ambulance stations in the UK was monitored. The resulting sample mean was 2846.6 and an unbiased estimate of the population variance was 312.4^2. Calculate a symmetric 90% confidence interval for the mean number of calls made to all ambulance stations in the UK in May 2000.

At the end of the year, the monthly figures for all the ambulance stations were obtained and the mean for May was found to lie outside the interval. How might this be explained?

4   The times, t minutes, taken by 18 children in an infant reception class to complete a jigsaw puzzle were measured. The results are summarised by $\sum t = 75.6$ and $\sum t^2 = 338.1$.

(a) Stating your assumptions, calculate a symmetric 95% confidence interval for the population mean time for children to complete the puzzle.

(b) The manufacturers of the puzzle indicate a mean completion time of 5 minutes. What conclusion might be made about the children in the class?

5   The acceleration due to gravity, g, is determined experimentally. In 5 independent determinations the values, in $\mathrm{m\,s}^{-2}$, are 9.79, 9.82, 9.80, 9.78, 9.84. It may be assumed that these values are observations from a normal distribution whose mean is g.

(a) Obtain a point estimate for g based on the 5 values.

(b) Calculate a symmetric 99% confidence interval for g, giving end-points to 3 decimal places.

(c) State whether or not, apart from rounding errors, the confidence interval found is exact.

6   A random sample of 12 fully-grown swallows (*Hirundo rustica*) were captured and released after their lengths, from tip of tail to tip of beak, were measured. The results, x cm, are summarised by $\sum x = 229.2$ and $\sum x^2 = 4389.16$.

(a) Assuming that the lengths are distributed normally, calculate a symmetric 98% confidence interval for the mean length of all fully-grown swallows.

(b) A thirteenth swallow was caught and found to have a length of 15.9 cm. Assuming that the mean and variance of the lengths of fully-grown swallows are approximated well by the sample estimates, is it likely that this swallow is fully grown?

7   The contents of 140 bags of flour selected randomly from a large batch delivered to a store are weighed and the results, w grams, summarised by $\sum(w - 500) = -266$ and $\sum(w - 500)^2 = 1178$.

(a) Calculate unbiased estimates of the batch mean and variance of the mass of flour in a bag.

(b) Calculate a symmetric 95% confidence interval for the batch mean mass.

The manager of the store believes that the confidence interval indicates a mean less than 500 g and considers the batch to be sub-standard. She has all of the bags in the batch weighed and finds that the batch mean mass is 501.1 g. How can this be reconciled with the confidence interval calculated in part (b)?

8   An environmental science student carried out a study of the incidence of lichens on a stone wall in Derbyshire. She selected, at random, 100 one-metre lengths of wall, all of the same height. The number of lichens in each section was counted and the results are summarised in the following frequency table.

Number of lichens	0	1	2	3	4	5	6
Number of sections	8	22	27	19	13	8	3

(a) Calculate the sample mean and an unbiased estimate of the population variance of the number of lichens per metre length of the wall.

(b) Calculate a symmetric 90% confidence interval for the mean number of lichens per metre length of the wall.

11.6   Confidence interval for a proportion

Many statistical investigations are concerned with finding the proportion of a population which has a specified attribute. Suppose you were a manufacturer of a version of an appliance designed for left-handed people. In order to assess the potential market you would be interested in the proportion of left-handed people in the population. It would be impossible to ask everybody and so you would have to rely on a sample. Suppose that you were able to obtain information from a random sample of 500 people and you found that 60 of them were left-handed. It would seem reasonable to estimate that the proportion, p, of the population who are left-handed is $\frac{60}{500} = 0.12$, or 12%. However, you need to be certain that this method gives you an unbiased estimate of p.

Consider the more general situation where a random sample of n people are questioned. Provided that n is much smaller than the population size the distribution of X, the number of people in a sample of size n who are left-handed, will be $B(n, p)$ since

- there are a fixed number of trials (n people asked)
- each trial has two possible outcomes (left-handed or right-handed)
- the outcomes are mutually exclusive (assuming that no one is ambidextrous)
- the probability of a person being left-handed is constant
- the trials are independent.

Let P be the random variable 'the proportion of people in a sample of size n who are left-handed'. Then $P = \frac{X}{n}$. The expected value of P is

$$E(P) = E\left(\frac{X}{n}\right) = \frac{1}{n}E(X)$$

$$= \frac{1}{n} \times np \qquad \text{(since the mean of a binomial distribution is } np\text{)}$$

$$= p.$$

Thus the proportion in the sample does provide an unbiased point estimate of the population proportion.

It would be more useful, however, to find a confidence interval for p since this gives an idea of the precision of the estimate. In order to do this you need to consider the distribution of P. For a large sample, X will be distributed approximately normally (Section 6.4) and so P will also be distributed approximately normally (see Section 4.1). In order to calculate a confidence interval you also need the variance of P. This can be found as follows.

$$\text{Var}(P) = \text{Var}\left(\frac{X}{n}\right)$$

$$= \frac{1}{n^2} \times \text{Var}(X)$$

$$= \frac{1}{n^2} \times npq \qquad \text{(since the variance of a binomial distribution is } npq\text{)}$$

$$= \frac{pq}{n}.$$

If $X \sim B(n, p)$, then the sample proportion P, where $P = \frac{X}{n}$, is distributed approximately as $N\left(p, \frac{pq}{n}\right)$.

This approximation may be used when n is large enough that $np > 10$ and $nq > 10$. In practice, this will be achieved if there are more than 10 successes and 10 failures in n trials.

In Section 11.3 you saw that for the sampling distribution $\bar{X} \sim N\left(\mu, \frac{\sigma^2}{n}\right)$ a confidence interval for the mean is given by $\left[\bar{x} - z\frac{\sigma}{\sqrt{n}}, \bar{x} + z\frac{\sigma}{\sqrt{n}}\right]$. By analogy a confidence interval for the proportion is found by replacing $\bar{x}$ by p and $\frac{\sigma}{\sqrt{n}}$ by $\sqrt{\frac{pq}{n}}$ to give an approximate confidence interval $\left[p - z\sqrt{\frac{pq}{n}}, p + z\sqrt{\frac{pq}{n}}\right]$, where values of z are obtained as before.

This confidence interval is expressed in terms of p and q, which are not known (otherwise a confidence interval would not be required!). However, again similar to finding a confidence interval for the mean, these unknown quantities can be replaced by their estimates from the sample provided that the sample is large. The result is the interval

$$\left[\hat{p} - z\sqrt{\frac{\hat{p}\hat{q}}{n}}, \hat{p} + z\sqrt{\frac{\hat{p}\hat{q}}{n}}\right], \text{ where } \hat{p} = \frac{x}{n} \text{ and } \hat{q} = 1 - \hat{p} = 1 - \frac{x}{n}.$$

Given a large random sample, size n, from a population in which a proportion of members, p, has a particular attribute, an approximate $100(1-\alpha)\%$ confidence interval for p is

$$\left[\hat{p}-z\sqrt{\frac{\hat{p}\hat{q}}{n}},\hat{p}+z\sqrt{\frac{\hat{p}\hat{q}}{n}}\right],$$

where $\hat{p}$ is the sample proportion with this attribute, $\hat{q}=1-\hat{p}$ and the value of z is such that $P(|Z|\le z)=1-\alpha$, or $P(Z\le z)=1-\frac{1}{2}\alpha$.

This rule can now be used to calculate a confidence interval for the proportion of left-handers in the population. For the sample taken, $n=500$, $\hat{p}=0.12$ and $\hat{q}=0.88$, and for a 95% confidence interval, $z=1.96$. This gives a 95% confidence interval

$$\left[0.12-1.960\times\sqrt{\frac{0.12\times0.88}{500}},0.12+1.960\times\sqrt{\frac{0.12\times0.88}{500}}\right]=[0.092,0.148].$$

It is interesting to note that the answer does not depend on the size of the population, only the size of the sample. This will always be true if the sample size is much less than the population size so that the value of p is effectively constant. This fact can be used to calculate the sample size required to give a confidence interval of specified width, as illustrated in the following example.

Example 11.6.1
An opinion poll is to be carried out to estimate the proportion of the electorate of a country who will vote 'yes' in a forthcoming referendum. In a trial run a random sample of 100 people were questioned; 42 said they would vote 'yes'. Estimate the random sample size required to give a 99% confidence interval of the proportion with a width of 0.02.

For a 99% confidence interval, z takes the value 2.576, so the width of this interval is
$2\times2.576\sqrt{\frac{\hat{p}\hat{q}}{n}}$, where $\hat{p}$ and $\hat{q}$ are used to estimate p and q.

The trial run gives $\hat{p}=0.42$ and hence $\hat{q}=0.58$. Thus the required value of n is given by

$$2\times2.576\times\sqrt{\frac{0.42\times0.58}{n}}=0.02.$$

Rearranging and solving for n gives 16 165 to the nearest integer.

It is interesting to think that if this example referred to the UK, where the size of the electorate is about 44 million, then the sample required is only about 0.04% of the electorate. The problem lies in obtaining a random sample. In practice opinion polls do not rely on random samples but use sophisticated techniques which are meant to ensure representative samples.

Exercise 11C

1   In a study of computer usage a random sample of 200 private households in a particular town was selected and the number that own at least one computer was found to be 68. Calculate a symmetric 90% confidence interval of the percentage of households in the town that own at least one computer.

2   Of 500 cars passing under a road bridge on the M1 motorway 92 were found to be red.

   (a) Find a symmetric 98% confidence interval of the population proportion of red cars.

   (b) State any assumption required for the validity of the interval.

   (c) Describe a suitable population to which the interval applies.

   (d) If 50 students carried out this experiment at different places and times, what is the expected number of confidence intervals which would contain the population proportion of red cars?

3   A biased dice was thrown 600 times and resulted in 224 sixes. Calculate a symmetric 99% confidence interval of p, the probability of obtaining a six in a single throw of the dice.

   Estimate the smallest number of times the dice should be thrown for the width of the symmetric 99% confidence interval of p to be at most 0.08.

4   The board of trustees of a charitable trust wishes to make a change to the trust's constitution. In order for this to happen at least two-thirds of the members must vote for the change. Before the vote is taken, the secretary consults a random sample of 60 members and finds that 75% of them will vote for the change. Calculate a symmetric 95% confidence interval for the proportion of all members who will vote for the change.

   Nearer the time at which the vote is to be taken, the secretary consults a random sample of n members and finds, again, that 75% of them will vote for the change. Using this figure, he calculates a symmetric 99% confidence interval for the proportion of members who will vote for the change. This interval does not include the value two-thirds. Find the smallest possible value of n.   (OCR)

5   The compiler of crossword puzzles classifies a puzzle as 'easy' if 60% or more people attempting the puzzle can complete it correctly within 20 minutes. It is classified as 'hard' if fewer than 30% of people can complete it correctly within 20 minutes. All other puzzles are classified as 'average'. A particular puzzle was given to 150 competitors in a contest and 74 completed it correctly within 20 minutes. The compiler wishes to be 90% confident of correctly classifying the puzzle.

   (a) How should she classify the puzzle?

   (b) Can she be 95% confident that her classification is correct?

12 A survey of probability distributions

This chapter presents a summary of probability distributions and their properties, some of which you have met already. When you have completed it, you should

- know the main properties of a number of probability distributions, and when it is appropriate to use them.

12.1 Uniform distributions

If there is no reason to think that any outcome is more probable than any other, the distribution is said to be uniform. This may occur with either a discrete or a continuous random variable.

(i) The discrete uniform distribution $DU(n)$

In most examples of a discrete uniform distribution the sample space is a set of positive integers $\{1,2,3,\dots,n\}$ for some value of n. When you roll a fair dice the score has a uniform distribution with $n = 6$. When you draw a number out of a hat to find the winner of a raffle, it has a uniform distribution with n equal to the total number of tickets sold.

Since $P(X = x)$ is the same for each number x in the sample space, and the sum of all the probabilities is equal to 1, it follows that

$$P(X = x) = \frac{1}{n} \quad \text{for each } x \in \{1,2,3,\dots,n\}.$$

You can find the mean and variance of this distribution by using the sums

$$\sum_{x=1}^{n} x = \tfrac{1}{2}n(n+1) \quad \text{and} \quad \sum_{x=1}^{n} x^2 = \tfrac{1}{6}n(n+1)(2n+1),$$

which were derived in Higher Level Book 1 Section 2.3 and Higher Level Book 2 Section 1.6.

Then

$$\mu = E(X) = \sum_{x=1}^{n} x \times \frac{1}{n} = \frac{1}{n}\sum_{x=1}^{n} x$$

$$= \frac{1}{n} \times \tfrac{1}{2}n(n+1) = \tfrac{1}{2}(n+1),$$

and

$$\sigma^2 = \text{Var}(X) = \sum_{x=1}^{n} x^2 \times \frac{1}{n} - \mu^2$$

$$= \frac{1}{n}\sum_{x=1}^{n} x^2 - \mu^2$$

$$= \frac{1}{n} \times \tfrac{1}{6}n(n+1)(2n+1) - \left(\tfrac{1}{2}(n+1)\right)^2$$

$$= \tfrac{1}{6}(n+1)(2n+1) - \tfrac{1}{4}(n+1)^2$$

$$= \tfrac{1}{12}(n+1)(2(2n+1) - 3(n+1)) = \tfrac{1}{12}(n+1)(n-1) = \tfrac{1}{12}(n^2 - 1).$$

For other sample spaces you can find the mean and variance by using the algebra of expectations. For example, if the sample space consists of the 11 even numbers from 60 to 80, the uniform random variable Y can be expressed in terms of $X \sim D(11)$ by $Y = 58 + 2X$, so

$$E(Y) = E(58 + 2X) \qquad \text{and} \qquad \text{Var}(Y) = \text{Var}(58 + 2X)$$
$$= 58 + 2E(X) \qquad\qquad\qquad\qquad = 2^2 \times \text{Var}(X)$$
$$= 58 + 2 \times \tfrac{1}{2}(11+1) = 70 \qquad\qquad = 4 \times \tfrac{1}{12}\left(11^2 - 1\right) = 40.$$

(ii)   The continuous uniform distribution $U(a,b)$

A familiar example of this is the error when people give their age in completed years. This may be anything between 0 and 1 years, and for a randomly selected individual there is no reason to prefer any value to any other. Denoting the error by r, the random variable R has probability density

$$f(r) = \begin{cases} 1 & \text{for } 0 \le r < 1, \\ 0 & \text{otherwise.} \end{cases}$$

From this it is easy to calculate the mean and variance

$$E(R) = \int_0^1 r \times 1 \, dr = \left[\tfrac{1}{2}r^2\right]_0^1 = \tfrac{1}{2},$$

$$\text{Var}(R) = \int_0^1 r^2 \times 1 \, dr - (E(R))^2$$
$$= \left[\tfrac{1}{3}r^3\right]_0^1 - \left(\tfrac{1}{2}\right)^2 = \tfrac{1}{3} - \tfrac{1}{4} = \tfrac{1}{12}.$$

More generally, a random variable X may take any value in the interval $[a,b]$. If the distribution is uniform, the probability density graph is a horizontal line segment. For the area under this graph to equal 1, the probability density must be given by the equation

$$f(x) = \begin{cases} \dfrac{1}{b-a} & \text{for } a \le x < b, \\ 0 & \text{otherwise.} \end{cases}$$

You can find the mean and variance for the random variable X from that for R. Since $X = a$ corresponds to $R = 0$, and $X = b$ to $R = 1$, these are connected by the equation

$$X = a + (b-a)R.$$

So

$$\mu = E(X) \qquad\qquad \text{and} \qquad \sigma^2 = \text{Var}(X)$$
$$= a + (b-a)E(R) \qquad\qquad\qquad = (b-a)^2 \text{Var}(R)$$
$$= a + (b-a) \times \tfrac{1}{2} = \tfrac{1}{2}(a+b) \qquad = \tfrac{1}{12}(b-a)^2.$$

12.2 Sequences of independent Bernoulli trials

The basis of this set of probability models is the Bernoulli trial, a single trial with two possible outcomes, 'success' or 'failure', with probabilities p and q, where $p + q = 1$. If you carry out a sequence of independent Bernoulli trials, various questions can be asked:

How many successes will there be in a sequence of n trials?
How many trials must you carry out until you get a success?
How many trials must you carry out until you get r successes?

The answers to these questions are given by the binomial, geometric and negative binomial probability distributions. You have already met the first two of these in Higher Level Book 1 Chapter 34 and in Chapter 2 of this book.

(i)   The Bernoulli probability distribution $B(1, p)$

The random variable X used to describe this distribution is the number of successes in the trial. This can take either of two values: 0 if the outcome is failure, 1 if success. So the sample space is the set $\{0,1\}$, with the probability distribution shown in Table 12.1. It is proved in Section 6.4 that $\mu = p$ and $\sigma^2 = pq$.

x	0	1
$P(X = x)$	q	p

Table 12.1

(ii)   The binomial probability distribution $B(n, p)$

The random variable X is now the total number of successes in a sequence of n trials. This can be any number between 0 and n, so the sample space is the set $\{0,1,2,\ldots,n\}$ (that is, $x \in \mathbb{N}, x \le n$).

You need to find the probability that the sequence of n trials results in x successes and $n - x$ failures. To do this, consider the whole set of trials $\{T_1, T_2, \ldots, T_n\}$ and split it into two subsets, one of x trials and the other of $n - x$ trials. The number of ways of doing this is ^nC_x (see Higher Level Book 1 Section 3.6). For each of these, the probability that all the trials in the first subset have successful outcomes is p^x, and the probability that all the trials in the second subset result in failure is q^{n-x}. So, in the complete sequence of trials, the probability of x successes and $n - x$ failures is $^nC_x q^{n-x} p^x$. And since ^nC_x is equal to the binomial coefficient $\binom{n}{x}$ (Higher Level Book 1 Section 4.3),

$$P(X = x) = \binom{n}{x} q^{n-x} p^x \qquad \text{for } 0 \le x \le n.$$

The reason for the name 'binomial' is that this probability is the term involving p^x when $(q + p)^n$ is expanded by the binomial theorem. Notice that, since $q + p = 1$, the sum of all these probabilities, $\sum_{x=0}^{n} P(X = x)$, is equal to $1^n = 1$, as you would expect.

A simple way of finding the mean and variance for this distribution is to let $X_1, X_2, \ldots, X_n$ be the number of successes (0 or 1) in the n independent Bernoulli trials $T_1, T_2, \ldots, T_n$. For each of these trials

T_i you know that $E(X_i) = p$ and $\text{Var}(X_i) = pq$, and clearly the total number of successes is $X = X_1 + X_2 + \ldots + X_n$. So

$$\begin{aligned}\mu &= E(X) \\ &= E(X_1) + E(X_2) + \ldots + E(X_n) \\ &= p + p + \ldots + p \\ &= np\end{aligned}$$

and

$$\begin{aligned}\sigma^2 &= \text{Var}(X) \\ &= \text{Var}(X_1) + \text{Var}(X_2) + \ldots + \text{Var}(X_n) \\ &= pq + pq + \ldots + pq \\ &= npq.\end{aligned}$$

(iii) The geometric probability distribution $\text{Geo}(p)$

When you roll a dice you hope that you won't have to wait too long to get the first six, but you know from experience that you may be unlucky. Theoretically the random variable X, the number of the trials needed to achieve the first success, may take any of the values 1, 2, 3,... without limit, so the sample space is the complete set of positive integers $\mathbb{Z}^+$.

If X is equal to x, this means that the outcome of the nth trial is success, but the outcomes of the previous $x - 1$ trials are all failure. The probability that this occurs is therefore pq^{x-1}. That is,

$$P(X = x) = pq^{x-1} \quad \text{for } x \in \mathbb{Z}^+.$$

This is the geometric probability distribution. It is shown in the Appendix Section A.2 that

$$\mu = \frac{1}{p} \quad \text{and} \quad \sigma^2 = \frac{q}{p^2}.$$

Another useful result is that the cumulative distribution function for $\text{Geo}(p)$ is

$$P(X \le x) = 1 - q^x = 1 - (1 - p)^x \qquad \text{(see Section 2.1).}$$

(iv) The negative binomial probability distribution $\text{NB}(r, p)$

Just as you can generalise from Bernoulli probability (1 trial) to binomial probability (n trials), you can generalise from geometric probability (the number of trials for the first success) to 'negative binomial' probability (the number of trials for r successes).

The random variable X now denotes the number of trials that you have to carry out so as to achieve r successes. This number must of course be at least r, so the sample space is the set $\{r, r+1, r+2, \ldots\}$ (that is, $x \in \mathbb{Z}^+, x \ge r$).

To find $P(X = x)$, note first that the outcome of the final, xth, trial has to be a success, and that the outcomes of the previous $x - 1$ trials have to include $r - 1$ successes. The number of failures in these $r - 1$ trials must therefore be

$$(x - 1) - (r - 1) = x - r.$$

By an argument similar to that used in subsection (ii) above for finding binomial probabilities, but with $x - 1$ in place of n and $r - 1$ in place of x,

$$\begin{aligned}P(X = x) &= p \times \binom{x-1}{r-1} q^{x-r} p^{r-1} \\ &= \binom{x-1}{r-1} q^{x-r} p^r \qquad \text{for } x \ge r.\end{aligned}$$

The reason for the name 'negative binomial' is given in the Appendix Q2.7, where it is shown that $\binom{x-1}{r-1} q^{x-r}$ is the expression for successive terms of the negative binomial expansion of $(1-q)^{-r}$. It follows that the sum of the probabilities $P(X = x)$ over the sample space is

$$\sum_{x=r}^{\infty} \binom{x-1}{r-1} q^{x-r} p^r = (1-q)^{-r} p^r$$
$$= p^{-r} \times p^r = 1,$$

as you would expect. Geometric probability is the special case of negative binomial probability with $r = 1$, so $\text{Geo}(p)$ could also be written as $\text{NB}(1, p)$.

To find the mean and variance you can use an argument like that used for binomial probability in subsection (ii) above, but with the geometric distribution in place of the Bernoulli distribution. You can think of the process of continuing trials up to the rth success as made up of r successive independent sequences of trials up to the first success. If the random variables for these r identical geometric distributions are denoted by $X_1, X_2, \ldots, X_r$, then the total number of trials is

$$X = X_1 + X_2 + \ldots + X_r.$$

And since, for each of these geometric trials $\text{E}(X_i) = \frac{1}{p}$ and $\text{Var}(X_i) = \frac{q}{p^2}$ (see subsection (iii)), the mean and variance for the negative binomial distribution are

$$\mu = \text{E}(X) = \text{E}(X_1) + \ldots + \text{E}(X_r) \qquad \text{and} \qquad \sigma^2 = \text{Var}(X) = \text{Var}(X_1) + \ldots + \text{Var}(X_r)$$
$$= \frac{1}{p} + \frac{1}{p} + \ldots + \frac{1}{p} \qquad\qquad\qquad = \frac{q}{p^2} + \frac{q}{p^2} + \ldots + \frac{q}{p^2}$$
$$= \frac{r}{p} \qquad\qquad\qquad\qquad\qquad\qquad = \frac{rq}{p^2}.$$

12.3 Sampling without replacement

Suppose you have a bag containing N discs, M of which are green and $N - M$ red. If you draw one disc out of the bag, the probability that it will be green is equal to the proportion of green discs in the bag. So this is a single Bernoulli trial, with $p = \frac{M}{N}$.

Now suppose that you repeat this trial n times. What is the probability that x of them will be green and $n - x$ red?

The answer is that it depends on how you carry out the trials. If after each trial you put the disc you have drawn back in the bag, then each trial will be conducted on exactly the same terms, with N discs of which M are green. So the distribution will be binomial, $\text{B}\left(n, \frac{M}{N}\right)$. This is called sampling with replacement.

But if you don't put the discs you have drawn back in the bag, the probability distribution for the first draw will be $\text{B}\left(1, \frac{M}{N}\right)$, but the probabilities for successive draws will depend on what has happened so far. For example, if the first disc drawn is green, then before the second draw there will be $N-1$ discs in the bag of which $M-1$ are green; but if the first disc is red, then there will be $N-1$ discs in the bag of which M are green. So the probability distribution for the second trial will be $B\left(1, \frac{M-1}{N-1}\right)$ or

$B\left(1, \frac{M}{N-1}\right)$ depending on whether the first disc drawn is green or red. This is **sampling without**

replacement. The conditions for binomial probability are not satisfied, and a different probability model is needed. This is called the hypergeometric probability model – a rather misleading name, since it is more closely related to binomial probability than to geometric probability.

The hypergeometric probability distribution $\mathrm{Hyp}(n,M,N)$

As with the binomial distribution, the random variable X is the number of successes (that is, green discs) in a sequence of n trials. So the sample space is again $\{0,1,2,\ldots,n\}$.

To find the probability of a sample with x successes and $n-x$ failures you can use a counting argument. Fig. 12.2 illustrates the situation when such a sample has been drawn. Out of the M original green discs x have been selected, and out of the $N-M$ red discs $n-x$ have been selected. The numbers of ways of doing this are respectively $^{M}\mathrm{C}_{x}$ and $^{N-M}\mathrm{C}_{n-x}$, so the total number of possible samples of this kind is $^{M}\mathrm{C}_{x} \times ^{N-M}\mathrm{C}_{n-x}$. But the total number of different samples of size n that can be drawn out of the bag is $^{N}\mathrm{C}_{n}$.

So $\quad P(X = x) = \dfrac{^{M}\mathrm{C}_{x} \times ^{N-M}\mathrm{C}_{n-x}}{^{N}\mathrm{C}_{n}}$

$\qquad = \dfrac{\dbinom{M}{x} \times \dbinom{N-M}{n-x}}{\dbinom{N}{n}} \quad$ for $0 \le x \le n.$

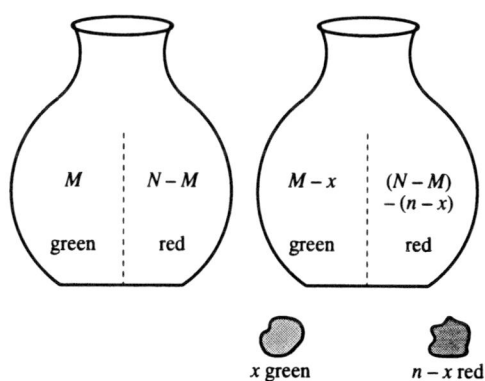

Fig. 12.2

Surprisingly the mean of this distribution is the same as that of the corresponding binomial distribution for sampling with replacement, but the variance is slightly smaller. The formulae are

$$\mu = np \qquad \text{and} \qquad \sigma^2 = npq \times \frac{N-n}{N-1},$$

where $p = \dfrac{M}{N}$ and $q = 1 - p$. A method of proving these results is described in the Appendix Section A.5.

12.4 Random events in time

The probability models in this section apply to sequences of events which occur randomly in time, such as the emission of particles from a radioactive source or emergency telephone calls to a fire station. Although the times when these occur are irregular and unpredictable, it will be possible to identify an average rate of occurrence. If the events occur

    singly and independently of each other
    at a constant average rate

then you can ask questions such as:

How many of the events will occur in one unit of time?
How long will it be before the first event occurs?

Answers to these questions are provided by the Poisson and exponential probability distributions. You have already met these in Higher Level Book 1 Chapter 35 and in Chapter 2 of this book.

Both these probability distributions depend on a single parameter m, the average rate of occurrence of the event per unit time.

(i)   The Poisson probability distribution $\text{Po}(m)$

The random variable X denotes the number of times that the event occurs in a unit of time. This is a discrete random variable with sample space $\mathbb{N}$, with probability function

$$P(X = x) = e^{-m} \frac{m^x}{x!}.$$

To show that the sum of these probabilities is equal to 1, and to find the mean and variance, you need to use the exponential series (see the Appendix Section A.3). You can then prove that

$$\mu = m \quad \text{and} \quad \sigma^2 = m.$$

The fact that the mean and variance are equal is a useful pointer to whether or not a Poisson probability model applies in a particular situation. For example, suppose that you count the number of cars passing a checkpoint in successive minutes over a period of three hours. You could then make an estimate of the mean and variance from the data. If these were close to each other, then it would be worth investigating whether the distribution of the data resembled a Poisson distribution. (How to do this is explained in Chapter 13.) But if the estimated mean and variance were very different, you could conclude at once that the conditions for Poisson probability did not apply to this experiment.

An important property of Poisson probability is that, if two random sequences of events are taking place in parallel, with occurrences modelled by Poisson probabilities $P(a)$ and $P(b)$, then the occurrence of one or other event is modelled by Poisson probability $P(a + b)$.

(ii)   The exponential probability distribution $\text{Exp}(m)$

Let X denote the time that you have to wait until the first event occurs. This time must be positive, but it need not of course be an integer. The time can be any positive real number, so X is a continuous random variable and the sample space is $\mathbb{R}^+$. The probability density function for X is

$$f(x) = \begin{cases} me^{-mx} & \text{for } x \geq 0, \\ 0 & \text{otherwise,} \end{cases}$$

and the cumulative distribution function is

$$F(x) = \begin{cases} 1 - e^{-mx} & \text{for } x \geq 0, \\ 0 & \text{otherwise.} \end{cases}$$

The mean and variance are

$$\mu = \frac{1}{m} \quad \text{and} \quad \sigma^2 = \frac{1}{m^2}. \quad \text{(See the Appendix Section A.6.)}$$

(iii)   A combined equation for Poisson and exponential probability

If the average rate of occurrence in unit time is m, the average rate of occurrence in x units of time is mx. The probability that the event occurs n times in x units of time is then

$$e^{-mx} \frac{(mx)^n}{n!}.$$

From this expression you can obtain the equations for both Poisson and exponential probability. If you put x equal to 1, you get $e^{-m} \dfrac{m^n}{n!}$, the Poisson probability of n occurrences in unit time. And if you put $n = 0$, you get e^{-mx}, the probability of no occurrences in time x, from which you can deduce the cumulative probability that the event first occurs within the first x units of time,

$$P(X \le x) = 1 - e^{-mx}.$$

12.5   Normal probability

Normal probability was introduced in Higher Level Book 2 Chapter 11. It is important both for its special theoretical properties and also because it provides a good approximate model in many practical situations where measured quantities are subject to random errors. You might think that this is rather surprising, since the sample space for normal probability density is the complete set $\mathbb{R}$ of real numbers, but it is used to model measurements which are restricted to a finite interval. For example, the heights of adult Danish women might be modelled by normal probability with mean 1.64 metres and variance 0.01 metres^2; but you will never meet one with height less than 0 metres or greater than 3 metres. The justification for using a normal model is that, for all practical purposes, the probability of getting values of the random variable more than a few standard deviations away from the mean is negligibly small.

(i)   The standardised normal probability distribution $N(0,1)$

It is simplest to begin with $N(0,1)$. You know that it is usual to denote the probability density function by the letter ϕ, so that its equation is

$$\phi(x) = \frac{1}{\sqrt{2\pi}} e^{-\frac{1}{2}x^2}.$$

The reason for the factor $\dfrac{1}{\sqrt{2\pi}}$ is so that the area under the complete graph, given by $\displaystyle\int_{-\infty}^{\infty} \phi(x)\,dx$, is equal to 1. This is quite difficult to prove, but you can verify it numerically with a calculator. The calculator cannot cope with infinite limits of integration, but you should find that if you calculate

$$\int_{-10}^{10} \frac{1}{\sqrt{2\pi}} e^{-\frac{1}{2}x^2}\,dx$$

you get an answer which is as close to 1 as the accuracy of the calculator allows.

Since x appears in the expression for $\phi(x)$ only in the exponent $-\frac{1}{2}x^2$, $\phi(x)$ is an even function. The mean of the distribution is therefore 0. You will find a method of proving that the variance is 1 in the Appendix Section A.4.

(ii)   The general normal probability distribution $N(\mu,\sigma^2)$

To get the most general form for the normal probability function, the first step is to stretch the probability density graph in the x-direction by a factor σ. To keep the total area under the graph equal to 1, there must be a corresponding stretch in the y-direction of factor $\dfrac{1}{\sigma}$. This gives the probability density for $N(0,\sigma^2)$ in the form

$$\frac{1}{\sigma}\phi\left(\frac{x}{\sigma}\right) = \frac{1}{\sigma\sqrt{2\pi}}e^{-\frac{1}{2}x^2/\sigma^2}.$$

A translation of μ in the x-direction then converts this to $N(\mu,\sigma^2)$, with probability density

$$\frac{1}{\sigma}\phi\left(\frac{x-\mu}{\sigma}\right) = \frac{1}{\sigma\sqrt{2\pi}}e^{-\frac{1}{2}(x-\mu)^2/\sigma^2}.$$

The random variable for $N(\mu,\sigma^2)$ is then $\sigma X + \mu$, where $X \sim N(0,1)$, so that

$$\begin{aligned}E(\sigma X + \mu) &= \sigma E(X) + \mu \\ &= \sigma \times 0 + \mu = \mu\end{aligned}$$

and

$$\begin{aligned}\text{Var}(\sigma X + \mu) &= \text{Var}(\sigma X) \\ &= \sigma^2\text{Var}(X) = \sigma^2.\end{aligned}$$

Important properties of normal probability are that, if X is a normal random variable and a, b are constants, then $aX + b$ is also a normal random variable; and, if X and Y are normal random variables, then so is $X + Y$.

Exercise 12

1   Draw graphs of the probability functions, and the corresponding cumulative distribution functions. (Use a calculator where it would help.)

(a) $DU(5)$ (b) $B(1,0.4)$ (c) $B(10,0.4)$ (d) $Geo(0.4)$

(e) $NB(3,0.4)$ (f) $Hyp(3,4,10)$ (g) $Po(2.4)$

2   Find the mean and variance of the probability functions in Question 1.

3   Use a calculator to display probability density graphs for the following distributions.

(a) $Exp(2.4)$ (b) $N(1,4)$ (c) $N(4,1)$

4   Use (a) graphs (b) algebra to explain why the variance of $U(1,n)$ must be less than that of $DU(n)$ for any integer $n > 1$.

5 The school winter term lasts for ten 5-day weeks. The cook serves chips on average once a week. To decide when to do so, he uses a random number program; chips are served on days when it produces either a 0 or a 9.

(a) Find the probability that during the term there will be three weeks in which chips are served more than once.

(b) In how many weeks of the term would you expect chips to be served at least once?

(c) Find the probability that chips will first be served on the second Wednesday of term.

(d) Find the probability that chips will be served for the tenth time on the last day of term.

6 When Alice and Zebedee play each other at chess, Alice wins three times as often as Zebedee. One year they decide to play one game every Saturday, 52 games in all.

(a) What is the probability that Alice will win exactly 39 times?

(b) Find the probability that Alice will win more than 40 times.

(c) Find the probability that Zebedee will first win on the sixth Saturday.

(d) How many games will they expect to play until Zebedee wins for the first time?

(e) What is the probability that Zebedee will win for the third time on the twelfth Saturday?

(f) Find the expectation and the standard deviation of the number of games they will play until Alice has her 30th win.

7 A box of chocolates contains 20 truffles and 30 caramels, but they all look the same. At a party of 12 people everyone takes a chocolate out of the box.

(a) Find the probability that four people get a truffle.

(b) What are the expectation and the standard deviation of the number of people who get a truffle?

8 In a card game a pile of 20 playing cards is placed face downwards on the table. Of these, six are picture cards. After shuffling one player takes the top four cards.

(a) Find the exact probability distribution of the number of picture cards she takes.

(b) Calculate the mean and variance of this distribution.

(c) Use the formulae for μ and σ^2 in Section 12.3 to check your answers to part (b).

9 A street is home to 100 voters, 30 of whom propose to vote for the Grey Party. Two statistics students each plan to investigate voting intentions by selecting 20 names at random from the list of voters, and asking each how they propose to vote. Their plans differ in one respect: if the same name comes up twice (or more), Abe proposes to count it twice, but Bea proposes only to count it once and to select another name in its place. What difference will this make to the mean and variance of the number of Grey voters they expect to get in their samples?

10 Cars travel at random intervals along a road. On average there are 3 cars going east and 2 going west each minute.

(a) It is safe to cross the road if no cars will come in either direction in the next 15 seconds. What is the probability of being able to cross the road when you want to?

(b) Find the probability that just one car will pass each way in the next half-minute.

(c) Find the probability that not more than 3 cars will pass in the next minute.

11   Random events occur independently in continuous time at an average rate of 6 per minute. Find, in seconds, the mean and standard deviation of the time up to the tenth occurrence.

12   A hall is lit by two kinds of light bulb. On average one Supabulb fails once every 100 days, and one Megalite fails once every 50 days. Find the probability that

(a) a bulb will fail within the next 30 days,

(b) exactly 3 bulbs will fail in the next 100 days.

13*   With the notation of Section 2.6, let y_n be the probability that exactly n events have occurred up to the time x.

(a) Explain why, if $n \geq 1$, $y_n + \delta y_n = y_n \times (1 - m\delta x) + y_{n-1} \times m\delta x$.

(b) Use part (a) to obtain the differential equation $\dfrac{dy_n}{dx} = m(y_{n-1} - y_n)$.

(c) Explain why, when $x = 0$, $y_n = 0$ for $n \geq 1$.

(d) Show that the equations in parts (b) and (c) are satisfied by

$$y_n = e^{-mx} \frac{(mx)^n}{n!} \quad \text{and} \quad y_{n-1} = e^{-mx} \frac{(mx)^{n-1}}{(n-1)!}.$$

(e) Show that, when $x = 1$, y_n is the Poisson probability that n events occur in unit time.

13 Chi-squared tests

This chapter introduces a method of testing whether theory is supported by practice. When you have completed it, you should be able to

- fit a theoretical distribution to given data
- use a χ^2 test with the appropriate number of degrees of freedom to carry out the corresponding goodness of fit test
- use a χ^2 test with the appropriate number of degrees of freedom to test for independence in a contingency table.

13.1 Comparing observed and expected frequencies

Table 13.1 shows the results obtained when a dice was thrown 120 times.

Score	1	2	3	4	5	6
Frequency	26	13	21	25	18	17

Table 13.1

How would you set about deciding whether these results indicate that the dice is a fair one? A start would be to compare these observed frequencies with the theoretical frequencies which you would expect if the dice is fair. The theoretical frequency for each score is found by multiplying the total frequency by the corresponding probability. In this case, the probability of each possible score is $\frac{1}{6}$ and so all the theoretical frequencies are equal to $120 \times \frac{1}{6} = 20$. These theoretical frequencies are usually known as **expected frequencies** and the process of calculating them is called 'fitting a theoretical distribution'. The equation defining expected frequency is

Expected frequency of a value = total frequency × probability of that value.

In Table 13.2 the second and third columns give the observed frequencies and the expected frequencies for the results of the dice experiment, where the observed and expected frequencies are denoted by f_o and f_e respectively.

You can see that the agreement between the observed and expected frequencies is not exact. This is what you would expect since the frequencies for each score are random variables: they vary each time the experiment is carried out. What the expected frequency tells you is the mean (or expected value) of each of these random variables.

The fourth column of Table 13.2 gives the difference $f_o - f_e$ for each class. If the agreement between experiment and theory is a good one, then you would expect these differences to be small. You might think that you could measure the total discrepancy between the observed and expected frequencies by finding the total of this column, $\sum(f_o - f_e)$.

Score	f_o	f_e	$f_o - f_e$	$(f_o - f_e)^2/f_e$
1	26	20	6	1.8
2	13	20	-7	2.45
3	21	20	1	0.05
4	25	20	5	1.25
5	18	20	-2	0.20
6	17	20	-3	0.45
Total	120	120	0	$X^2 = 6.2$

Table 13.2

However, you can see from the table that this total is zero, even though the observed and expected frequencies do not all agree. In fact, this quantity will always equal zero (because the sums of the observed and expected frequencies are both equal to the total frequency) and so it is no use as a measure of discrepancy. This problem could be overcome by squaring the differences and calculating $\sum (f_o - f_e)^2$.

However, this measure of the discrepancy is still not satisfactory because it makes no allowance for the size of the difference relative to the size of the expected frequencies. For example, the difference between an observed frequency of 11 and expected frequency of 10 would give the same contribution to this sum as the difference between an observed frequency of 101 and expected frequency of 100, even though the percentage difference between the two values is 10% in the first instance and 1% in the second. The measure of discrepancy which is actually used is $\sum \dfrac{(f_o - f_e)^2}{f_e}$. It is given the symbol X^2. The square emphasises that this is always a positive quantity.

$$X^2 = \sum \frac{(f_o - f_e)^2}{f_e}$$

The statistic X^2 gives a measure of the **goodness of fit** of the model; that is, how well the observed and theoretical frequencies agree. For perfect agreement it takes the value zero, and it increases as the differences between the observed and expected frequencies increase. The value of X^2 for the dice data is calculated in the last column of Table 13.2 and is equal to 6.2.

If you repeated the experiment with the same dice, you would get different values of f_o and hence a different value of X^2. In other words, X^2 is also a random variable. To decide whether the value 6.2 indicates a significant difference between the observed and expected frequencies it is necessary to study the probability distribution of X^2. This distribution is different from any of the distributions which you have already met. It is approximately related to a new family of continuous distributions which is introduced in the next section.

13.2   The chi-squared (χ^2) family of distributions

Fig. 13.3 shows probability density graphs for some of the members of the χ^2 family of distributions.

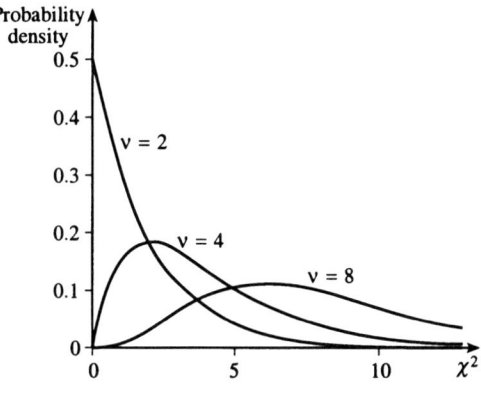

Fig. 13.3

The symbol χ is a Greek letter whose name 'chi' is pronounced as the 'ki' in 'kite'. Thus χ^2 is read as 'chi-squared'. The shape of each member of the family is determined by a single parameter, ν, which is known as the **number of degrees of freedom**. You write χ_ν^2 to indicate the member of the family with ν degrees of freedom. The mean and variance of a χ_ν^2 distribution are ν and 2ν respectively (see the Appendix Section A.4). This means that as ν increases the location of the distribution shifts to the right and the spread of the distribution increases. You can see this effect in Fig. 13.3, which shows χ_2^2, χ_4^2 and χ_8^2.

As with other distributions used for hypothesis testing, like the t-distribution, you are almost always interested in finding probabilities represented by the area under a χ^2 graph. You can find these directly by using the χ^2 cumulative distribution program on a calculator. You will usually want to find the probability that χ^2 is less or greater than a particular number. For example, $P\left(\chi_8^2 \le 10\right) = 0.735$, or $P\left(\chi_4^2 \ge 5\right) = 0.287$.

To decide which member of the χ^2 family gives an approximation to the distribution of X^2 you need to know the value of ν. As for the t-distribution, this is given by the number of independent values, in this case expected frequencies (or equivalently, classes), which are used in the calculation of X^2. There are 6 expected frequencies in Table 13.2, but they are not independent. Since the total of the expected frequencies must equal 120, there is 1 constraint. Thus

number of degrees of freedom = number of classes − number of constraints
$$= 6 - 1 = 5,$$

and the distribution of X^2 for this experiment can be approximated by the χ_5^2 distribution.

A useful way of thinking of a constraint in the context of χ^2 tests is as a piece of information which is obtained from the observed frequencies and then used in the calculation of the expected frequencies. In this example the total observed frequency of 120 was multiplied by the probabilities to find the expected frequencies. This gave one constraint. Later in this chapter you will see examples where there is more than one constraint.

To sum up the process so far:

Separate the sample space into a number of mutually exclusive classes.

Obtain n independent values of the random variable experimentally and assign each to its class.

Record the observed frequencies, f_o, in each class.

Calculate the expected frequencies, f_e, in each class from an appropriate probability model.

Then the measure of discrepancy $X^2 = \sum \dfrac{(f_o - f_e)^2}{f_e}$, where the sum is taken over all the classes, has an approximately χ_v^2 distribution, where v is the number of degrees of freedom, calculated as

$$v = \text{number of classes} - \text{number of constraints}.$$

Notice the word 'approximately'. The nature of the approximation is rather like using a normal distribution to approximate to binomial probability (see Section 6.4). You will remember that this approximation was very close provided that the expected numbers of successes and failures were not too small. Similarly the χ^2 approximation is very close provided that the frequencies in every class are not too small.

Most people who use statistics work to the rule that the classes should be chosen so that the expected frequency in any class is not smaller than 5. If it is, then that class should be combined with another to make a class for which the expected frequency is larger.

Classes should be chosen so that the expected frequency in each class is at least 5.

Before you go on to apply the procedure, it is worth noting a different way of calculating the measure of discrepancy, by writing

$$\frac{(f_o - f_e)^2}{f_e} = \frac{f_o^2 - 2f_o f_e + f_e^2}{f_e}$$

$$= \frac{f_o^2}{f_e} - 2f_o + f_e.$$

Now both $\sum f_o$ and $\sum f_e$ are equal to n. It follows that

$$\sum \frac{(f_o - f_e)^2}{f_e} = \sum \frac{f_o^2}{f_e} - 2n + n = \sum \frac{f_o^2}{f_e} - n.$$

It is often quicker to use this form to calculate the discrepancy.

13.3   Carrying out a χ^2 goodness of fit test

For the dice data it was found that $X^2 = 6.2$. This value is based on expected frequencies which assume that the dice is fair and so this assumption forms the null hypothesis. The alternative is that the dice is not fair:

H_0: the dice is fair; H_1: the dice is not fair.

As explained above the distribution of X^2 can be approximated by χ_5^2. If H_0 is true, then the expected and observed frequencies should be similar and so the value of X^2 will be low. A high value of X^2 is unlikely and would lead you to reject H_0. The rejection region thus takes the form $X^2 \geq c$ where the critical value, c, depends on the significance level.

Fig. 13.4 shows the χ_5^2 distribution with acceptance and rejection regions for a 5% significance level.

From the calculator $P\left(\chi_5^2 \geq 6.2\right) = 0.287$. There is a 28.7% probability of getting a discrepancy X^2 as large as 6.2 with a fair dice. Since this is much greater than 5%, $X^2 = 6.2$ is well within the acceptance region with a 5% significance level. There is no reason to think that the dice is unfair.

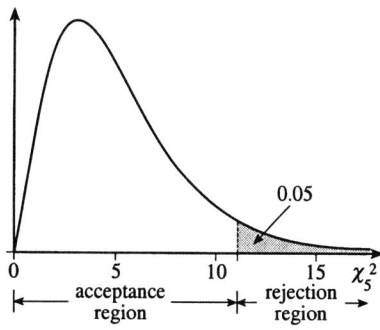

Fig. 13.4

The following example illustrates a χ^2 goodness of fit test in a situation in which the expected frequencies are not all the same.

Example 13.3.1

In genetic work it is predicted that the children with both parents of blood group AB will fall into blood groups AB, A and B in the ratio 2:1:1. Of a random sample of 100 such children 55 were blood group AB, 27 blood group A and 18 blood group B. Test at the 10% significance level whether the observed results agree with the theoretical prediction.

Take

H_0: the offspring fall into groups AB, A and B in the ratio 2:1:1;
H_1: the offspring do not fall into groups in the ratio 2:1:1.

The probabilities of falling into the groups AB, A and B are $\frac{1}{2}$, $\frac{1}{4}$ and $\frac{1}{4}$ respectively. The expected frequencies are found by multiplying these probabilities by the total frequency of 100, giving 50, 25 and 25. None of these is less than 5, so the distribution of X^2 can be approximated by a χ^2 distribution.

Table 13.5 sets out the information you need for the calculation of X^2.

Blood group	AB	A	B	Total
f_o	55	27	18	100
f_e	50	25	25	100
$f_o - f_e$	+5	+2	−7	0

Table 13.5

In this example there are 3 classes. There is 1 constraint because the total observed frequency of 100 was used in the calculation of the expected frequencies. Thus

v = number of classes − number of constraints = 3 − 1 = 2.

To find the discrepancy between the observed and the expected frequencies, calculate

$$\sum \frac{(f_o - f_e)^2}{f_e} = \frac{5^2}{50} + \frac{2^2}{25} + \frac{(-7)^2}{25} = 2.62.$$

Your calculator gives $P(\chi_2^2 \geq 2.62) = 0.270$, or 27.0%. This is greater than the 10% significance level, so 2.62 lies in the acceptance region. The null hypothesis is therefore accepted. The offspring fall into the groups AB, A and B in the ratio 2:1:1 as predicted by genetic theory.

13.4 Goodness of fit tests for discrete probability models

In the dice example in Section 13.1, the observed frequencies were compared with the expected frequencies obtained from the uniform probability distribution $DU(6)$. The question was, whether the experimental evidence is consistent with the hypothesis that the behaviour of the dice is described by this probability model.

The examples show how you can answer similar questions about other discrete probability models.

(i) Testing for a binomial model

Example 13.4.1

A sociologist doing research on family planning decides to investigate the gender make-up of large families. She selects 100 families at random from census data, each with 4 children under 16. Table 13.6 is her record of the girl/boy split in these families.

4g/0b	3g/1b	2g/2b	1g/3b	0g/4b	Total
11	27	23	31	8	100

Table 13.6

(a) Give reasons why a binomial distribution $B\left(4,\frac{1}{2}\right)$ might be an appropriate model.

(b) Test this hypothesis at significance levels of   (i) 5%,   (ii) 1%.

    (a)   In the population at large boys and girls occur more or less equally. Evidence suggests that, if one child is a boy (say), then the next child is no more likely to be a boy than a girl, or vice versa. So if the parents have decided in advance that they are going to have 4 children, and are happy to 'take what comes', then the basic requirements for a binomial probability model are satisfied.

        •   There is a fixed number of trials (in this case 4).

        •   The outcome of each trial is independent of the outcomes of other trials.

        •   The probability of 'success' on each trial is constant (in this case $\frac{1}{2}$).

    With these assumptions, a $B\left(4,\frac{1}{2}\right)$ distribution is appropriate.

(b)   The null and alternative hypotheses are

H_0: the gender split has a $B\left(4,\frac{1}{2}\right)$ distribution;

H_1: the gender split does not have a $B\left(4,\frac{1}{2}\right)$ distribution.

Under H_0, the probability of r boys and $4-r$ girls is

$$\binom{4}{r}\left(\tfrac{1}{2}\right)^{4-r}\left(\tfrac{1}{2}\right)^{r} = \binom{4}{r}\times\left(\tfrac{1}{2}\right)^{4} = \tfrac{1}{16}\times\binom{4}{r}.$$

The probabilities for $r = 0, 1, 2, 3, 4$ are therefore $\frac{1}{16}, \frac{4}{16} = \frac{1}{4}, \frac{6}{16} = \frac{3}{8}, \frac{4}{16} = \frac{1}{4}, \frac{1}{16}$. To get the expected frequencies these probabilities are multiplied by 100, which gives the values of f_e in Table 13.7.

g/b split	4g/0b	3g/1b	2g/2b	1g/3b	0g/4b	Total
f_o	11	27	23	31	8	100
f_e	6.25	25	37.5	25	6.25	100
$f_o - f_e$	4.75	2	−14.5	6	1.75	0

Table 13.7

Notice that the expected frequencies needn't be whole numbers, although the observed frequencies obviously are.

You can work out the discrepancy X^2 either as

$$\sum\frac{(f_o-f_e)^2}{f_e} = \frac{4.75^2}{6.25} + \frac{2^2}{25} + \frac{(-14.5)^2}{37.5} + \frac{6^2}{25} + \frac{1.75^2}{6.25}$$
$$= 3.61 + 0.16 + 5.6066\ldots + 1.44 + 0.49$$
$$= 11.3066\ldots$$

or as

$$\sum\frac{f_o^2}{f_e} - n = \left(\frac{11^2}{6.25} + \frac{27^2}{25} + \frac{23^2}{37.5} + \frac{31^2}{25} + \frac{8^2}{6.25}\right) - 100$$
$$= (19.36 + 29.16 + 14.1066\ldots + 38.44 + 10.24) - 100$$
$$= 111.3066\ldots - 100 = 11.3066\ldots \ .$$

Both methods have their advantages. The first draws attention to the entries which make the largest contribution to the discrepancy, which are those for 4 girls/0 boys and 2 girls/2 boys. This may suggest where to begin to look for flaws in the model. The second is easier to work out, since you don't need the last line in the table; you can go straight to the answer from the tabulated values of f_o and f_e.

Since there are 5 classes, with the single constraint that $\sum f_e = 100$, the number of degrees of freedom is $5-1 = 4$. The calculator then gives

$$P\left(X^2 \geq 11.3066\ldots\right) \approx P\left(\chi_4^2 \geq 11.3066\ldots\right)$$
$$= 0.0233\ldots \ ,$$

or 2.33%.

(i)   Since $2.33 < 5$, the value $X^2 = 11.31$ falls into the rejection region at the 5% significance level. You therefore reject H_0, and conclude that the data do not fit a $B\left(4,\frac{1}{2}\right)$ distribution.

(ii)   Since $2.33 > 1$, the value $X^2 = 11.31$ falls into the acceptance region at the 1% significance level. You therefore accept H_0, and conclude that the data are consistent with the hypothesis of a $B\left(4,\frac{1}{2}\right)$ distribution.

So the sociologist's conclusion depends on the level of significance she chooses. The evidence for rejecting the $B\left(4,\frac{1}{2}\right)$ hypothesis is strong, but not very strong.

In some cases a binomial distribution may appear to be a suitable model, but a value of p is not known at the start. In this case it can be estimated from the data. The mean, μ, of a binomial distribution is given by $\mu = np$. Hence $p = \dfrac{\mu}{n}$. Taking $\bar{x}$ as an estimate of μ gives $p = \dfrac{\bar{x}}{n}$.

> In testing the goodness of fit of a binomial distribution, where p is unknown at the start, an estimate of p is calculated from the data as
>
> $$p = \frac{\bar{x}}{n}.$$

Example 13.4.2

In routine tests of germination rates, parsley seeds are planted in rows of 5 and the number of seeds which have germinated in each row after a fixed time interval is counted. Table 13.8 shows the results for 100 such rows.

Number of seeds germinated (x)	0	1	2	3	4	5
Number of rows f_o	0	0	8	23	43	26

Table 13.8

(a) Use the data to estimate a value for p, the probability that a seed germinates.

(b) Calculate the expected frequencies for the model $B(5, p)$.

(c) Use a χ^2 goodness of fit test at the 5% significance level to test the suitability of the model $B(5, p)$.

(a)   For the data in Table 13.8, $\sum xf_o = 387$, so $\bar{x} = 3.87$. Thus the estimate of p is $\frac{1}{5} \times 3.87 = 0.774$.

(b)   The null and alternative hypotheses are

   H_0: the data can be modelled by $B(5, 0.774)$;
   H_1: the data cannot be modelled by $B(5, 0.774)$.

The expected frequencies are found by multiplying the binomial probabilities by the total observed frequency of 100. You can calculate the binomial probabilities from the binomial formula $\dbinom{5}{r} \times (0.226)^{5-r} \times (0.774)^r$, but with awkward numbers like these you may prefer to use the binomial probability program on your calculator. The values of f_e are given in the bottom row of Table 13.9, correct to 2 decimal places.

Number germinated (x)	0	1	2	3	4	5	Total
f_o	0	0	8	23	43	26	100
f_e	0.06	1.01	6.92	23.68	40.55	27.78	100

Table 13.9

Another step is needed before you go on to calculate v. You will notice that two of the values of f_e are less than 5, so the classes for $x = 0$, $x = 1$ and $x = 2$ should be combined to produce a single class with a large enough expected frequency. This is shown in Table 13.10.

Number germinated (x)	0, 1 or 2	3	4	5	Total
f_o	8	23	43	26	100
f_e	7.99	23.68	40.55	27.78	100

Table 13.10

The number of classes after combination, in this case 4, is used in the calculation of the degrees of freedom. There are 2 constraints because two pieces of information have been obtained from the observed frequencies to calculate the expected frequencies: these are the total frequency (100) and the value of p (0.774). Thus $v = 4 - 2 = 2$.

From Table 13.10 you can calculate that $X^2 = 0.2816...$, and the calculator gives

$$P(X^2 \geq 0.2816...) \approx P(\chi_2^2 \geq 0.2816...) = 0.8686... , \text{ or } 86.9\%.$$

This is far larger than 5%, so the value $X^2 = 0.2816...$ is well within the acceptance region. That is, the binomial distribution $B(5, 0.774)$ is a very good model for the data.

If you combine classes when you do not have to, then you increase the probability of making a Type II error, that is keeping a false null hypothesis.

You may have noticed that the alternative hypothesis for a χ^2 goodness of fit test does not specify the way in which the null hypothesis might be incorrect. If the null hypothesis is rejected, it could be either because the value of the parameter which was used was incorrect or because the conditions for the distribution to be a suitable model are not met. This point is explored further in some of the exercises.

(ii) Testing for a geometric model
Your calculator probably has a program to create a sequence of random numbers. These can be used to pick winners in a lottery, or to provide a strategy in a game of chance, or to simulate values of a random variable in a probability model. People who make use of random numbers professionally take a lot of trouble to make sure that the sequence really is random.

How can you check a sequence for randomness?

One requirement is obviously that, in a long sequence, each number occurs roughly the same number of times. But this doesn't guarantee randomness by itself. For example, in the sequence

123 456 789 123 456 789 123 456 789 123 ...

the frequency of each of the digits from 1 to 9 is the same, but this is hardly a random sequence. So you need to back up the equal frequency test with other more sophisticated tests.

Here are the first 40 of a sequence of 200 digits between 1 and 9 generated by a calculator.

24137 16888 67756 68713 78376 15719 64469 26846 ...

One way of testing this for randomness is to ask, as you read along the sequence, how many digits you have to read until you reach the next multiple of 3. To answer this, rewrite it placing the gaps after each multiple of 3. This has the effect of breaking the sequence into 'runs' so that only the last digit in each run is a multiple of 3.

2413 716 8886 7756 6 8713 783 76 15719 6 446 9 26846 ...

The lengths of the runs are 4, 3, 4, 4, 1, 4, 3, 2, 5, 1, 3, 1, 5, ...

If the digits are random, the length of a run L is a random variable with a geometric distribution (see Section 2.1). If you consider a non-multiple of 3 as a 'failure', and a multiple of 3 as a 'success', then the probability of a success is $p = \frac{3}{9} = \frac{1}{3}$, so $L \sim \text{Geo}\left(\frac{1}{3}\right)$.

Example 13.4.3 investigates this for the complete sequence of 200 digits.

Example 13.4.3
In a sequence of 200 digits the length of runs up to the next multiple of 3 has the frequency distribution in Table 13.11. Test the sequence for randomness, at the 10% level, by finding the goodness of fit to $\text{Geo}\left(\frac{1}{3}\right)$.

Length of run	1	2	3	4	5	6	7	8	9	10	11	Total
f_o	15	14	14	10	5	2	0	1	1	1	1	64

Table 13.11

You may prefer to work this example using your own sequence of 200 digits instead of the one used to produce Table 13.11.

It is encouraging that the sequence of 200 digits split into 64 runs. Since the probability that any digit is a multiple of 3 is $\frac{3}{9} = \frac{1}{3}$, the expected number of multiples of 3 (and therefore of runs) in a sequence of 200 digits is $200 \times \frac{1}{3} = 66\frac{2}{3}$.

The null and alternative hypotheses are

H_0: the data can be modelled by $\text{Geo}\left(\frac{1}{3}\right)$;

H_1: the data cannot be modelled by $\text{Geo}\left(\frac{1}{3}\right)$.

In $\text{Geo}\left(\frac{1}{3}\right)$ the probabilities of runs of length $1,2,3,\ldots,x,\ldots$ are

$$\tfrac{1}{3},\tfrac{1}{3}\times\tfrac{2}{3}=\tfrac{2}{9},\tfrac{1}{3}\times\left(\tfrac{2}{3}\right)^{2}=\tfrac{4}{27},\ldots,\tfrac{1}{3}\times\left(\tfrac{2}{3}\right)^{x-1},\ldots.$$

Multiplying these by 64 gives the expected frequencies in Table 13.12.

Length of run	1	2	3	4	5	6	7	8	9	10	11	Total
f_o	15	14	14	10	5	2	0	1	1	1	1	64
f_e	21.33	14.22	9.48	6.32	4.21	2.81	1.87	1.25	0.83	0.55	0.37	…

Table 13.12

Clearly some classes must be combined to produce expected frequencies greater than 5. There are two possibilities:

- combine all the classes from 5 onwards
- combine the classes for 4 and 5, and all the classes from 6 onwards.

Table 13.13 is based on the second of these. Unlike the binomial distribution, there is no upper limit for the value of X, so the entry for '6 or more' is most simply calculated by subtracting the sum of the entries for 1, 2, 3, 4 and 5 from 64.

Length of run	1	2	3	4 or 5	6 or more	Total
f_o	15	14	14	15	6	64
f_e	21.33	14.22	9.48	10.53	8.44	64

Table 13.13

The calculation then proceeds in the usual way. There are 5 classes and 1 constraint, that $\sum f_e = 64$, so $v = 5-1 = 4$. From Table 13.13, $X^2 = 6.639\ldots$, and then

$$P\left(X^2 \geq 6.639\ldots\right) \approx P\left(\chi_4^2 \geq 6.639\ldots\right) = 0.156\ldots,$$

or 15.6%.

Since this is greater than the significance level of 10%, the null hypothesis is accepted: the data can be modelled by $\text{Geo}\left(\frac{1}{3}\right)$. So, on this criterion, the method used by the calculator to create sequences of random digits is satisfactory.

(iii) Testing for a Poisson model

The Poisson probability formula was introduced in Higher Level Book 1 Section 35.1 by using data relating to the number of calls arriving at a switchboard in time intervals of 5 minutes. These data are reproduced in Table 13.14.

Number of calls	0	1	2	3	4 or more
Frequency	71	23	4	2	0

Table 13.14

A χ^2 goodness of fit test can now be applied to these data. Use a significance level of 5%. The value of the parameter, m, the population mean, is not known and so it is estimated by the sample mean. You can check that the sample mean is 0.37. So the null and alternative hypotheses for a goodness of fit test are

H_0: the data can be modelled by $Po(0.37)$;
H_1: the data cannot be modelled by $Po(0.37)$.

The expected frequencies are found by multiplying the Poisson probabilities by the total observed frequency, 100. You can find these probabilities either from the formula $P(X = x) = e^{-0.37} \dfrac{0.37^x}{x!}$ or from the calculator program for Poisson probabilities. For example, the expected frequency for 3 calls is $e^{-0.37} \dfrac{0.37^3}{3!} \times 100 = 0.58$. You can check that you agree with the other expected frequencies in Table 13.15. As for the geometric distribution there is no theoretical upper limit on the number of calls and so the expected frequency for 4 or more is found by subtracting the sum of the expected frequencies for $x = 0, 1, 2$ and 3 from the total frequency of 100.

Number of calls	0	1	2	3	4 or more	Total
f_o	71	23	4	2	0	100
f_e	69.07	25.56	4.73	0.58	0.06	100

Table 13.15

In Table 13.16 the last three classes have been combined.

Number of calls	0	1	2 or more	Total
f_o	71	23	6	100
f_e	69.07	25.56	5.37	100

Table 13.16

There are 3 classes after combination, and 2 constraints, since the mean (0.37) and the total frequency (100) are both found from the observed values, so $v = 3 - 2 = 1$. From Table 13.16, $X^2 = 0.384...$, and

$$P(X^2 \geq 0.384...) \approx P(\chi_1^2 \geq 0.384...) = 0.535... ,$$

or 53.5%.

This is much larger than the significance level of 5%, so H_0 is accepted: $Po(0.37)$ is a suitable model for the distribution of calls at the switchboard.

If the value of the mean, m, is known or given, this value would be used in the calculation of the expected frequencies and the number of constraints would be 1 rather than 2.

Exercise 13A

1 A dice is rolled 60 times. The results are shown in the table.

Score	1	2	3	4	5	6
Frequency	10	11	9	6	10	14

Use a χ^2 test at the 5% significance level to test the hypotheses

H_0: the dice is fair; H_1: the dice is not fair.

2 In experiments on the breeding of flowers a researcher obtained 95 magenta flowers with a green stigma, 32 magenta flowers with a red stigma, 26 red flowers with a green stigma and 7 red flowers with a red stigma. Genetic theory predicts that flowers of these types should occur in the ratios 9:3:3:1. Carry out a χ^2 test at the 1% significance level to see if the experimental results are in line with the theory.

3 When a tetrahedral dice is thrown the number landing face down counts as the score. Four such dice are thrown 200 times and the numbers of fours obtained are shown.

Number of fours	0	1	2	3	4
Frequency	20	47	83	41	9

(a) Use a χ^2 test at the 5% level to test whether the dice are fair, that is that a $B\left(4, \frac{1}{4}\right)$ model is appropriate.

(b) Use the data to estimate a value for p, the probability that the score is 4. Test at the 5% level whether a $B(4, p)$ model is appropriate.

4 Sixty samples of size 6 are taken after a manufacturing process. The number of defective items in each sample is recorded and the results are shown in the table.

Number of defectives	0	1	2	3	4	5	6
Number of samples	8	24	14	11	1	1	1

Calculate the mean of the frequency distribution and hence find an estimate of p, the probability that an individual item is defective. Using your estimate calculate the expected frequencies corresponding to a binomial distribution and perform a χ^2 goodness of fit test at the 1% level of significance.

5 I wish to investigate the efficiency of the random number generating function on my new calculator. I program it to produce digits from 0 to 9 inclusive, and record the lengths of runs up to the appearance of the next even number. A 10% level is to be used.

The results obtained were the following.

Length of run	1	2	3	4	5	6	7 or more
Frequency	26	13	8	2	0	1	0

Carry out the test and state your conclusion.

6   Despite the result of the test in Question 5, I am not convinced of the randomness of the numbers generated. The digits generated while obtaining the 50 runs yielded the following probability distribution.

Digit	0	1	2	3	4	5	6	7	8	9
Frequency	11	13	5	7	12	6	6	7	6	17

Conduct a χ^2 goodness of fit test at the 10% level using the method of Section 13.1.

7   The number of phone calls I received each day during the first three months of 2006 are recorded in the table.

Number of calls	0	1	2	3	4
Number of days	44	24	14	6	2

It is suspected that they follow a Poisson distribution. Calculate the mean of the frequency distribution and use it to calculate the expected frequencies for $x = 0, 1, 2, 3, 4$ or more. Hence carry out a χ^2 test of goodness of fit at the 5% level.

8   The first 100 draws in the UK National Lottery produced the following results. (Six balls are drawn each time.)

Ball numbers	1–7	8–14	15–21	22–28	29–35	36–42	43–49
Frequency	88	80	80	91	87	78	96

Carry out a χ^2 goodness of fit test at the 5% level to test the null hypothesis
  $P(1–7) = P(8–14) = \ldots = P(43–49) = \frac{1}{7}$.

9   The results of Question 8 could also be grouped according to the colours of the balls used in the draws.

Colour	White	Blue	Pink	Green	Yellow
Ball numbers	1–9	10–19	20–29	30–39	40–49
Frequency	107	125	116	114	138

Carry out a χ^2 goodness of fit test at the 10% level to test the null hypothesis
  $P(W) = \frac{9}{49}$, $P(B) = P(P) = P(G) = P(Y) = \frac{10}{49}$.

10   Five drawing pins are thrown onto a table. The number landing point up is recorded. The experiment is conducted 50 times in all, resulting in the following frequency distribution.

Number landing point up	0	1	2	3	4	5
Frequency	3	9	20	15	3	0

Calculate the mean of the frequency distribution, estimate p, the probability of a pin landing point up, and carry out a goodness of fit test at the 5% significance level.

11 The goals scored in each match by a football team during two successive seasons are shown in the table.

Number of goals	0	1	2	3	4	5	6	7
Number of matches	14	32	21	12	3	1	0	1

It is suspected that a Poisson distribution is appropriate.

(a) In the season previous to these two seasons, the mean number of goals scored per match was 1. Carry out a χ^2 goodness of fit test of the model $Po(1)$. Use a 5% significance level.

(b) Calculate the mean of the frequency distribution in the table. Test whether a Poisson distribution with this mean gives a better fit to the data. Use a 5% significance level.

12 A survey is carried out at a supermarket till. When the till opens, the number of customers up to and including the first person to use one of the carrier bags provided by the supermarket is recorded. This is repeated on 100 consecutive days. The data are summarised in the table below.

Number of customers	1	2	3	4	>4
Frequency	79	15	3	3	0

It is thought that this distribution may be modelled by a geometric distribution with parameter p, where p is the probability that a person uses a supermarket carrier bag.

(a) Calculate the mean and hence obtain an estimate of p.

(b) Carry out a test, at the 5% significance level, of the goodness of fit of the model to the data.

13 The number of accidents occurring on a busy road each week is recorded for one year. The results are shown in the table.

Number of accidents (x)	0	1	2	3	4	5
Number of weeks	26	12	10	3	1	0

It is suspected that a Poisson distribution is appropriate. Calculate $\bar{x}$, and use it to find the expected frequencies for $x = 0, 1, 2, 3$ or more. Conduct a χ^2 goodness of fit test at the 2.5% level of significance.

13.5 Testing the goodness of fit of a normal model

So far the goodness of fit tests which have been carried out have been applied to models for discrete variables. It is also possible to apply the method to continuous data which have been grouped into suitable class intervals. Here is an example of testing the goodness of fit of the normal distribution.

Example 13.5.1

The height, in centimetres, gained by a conifer in its first year after planting is denoted by the random variable H. The value of H is measured for a random sample of 86 conifers and the results obtained are summarised in Table 13.17.

H	< 35	35–45	45–55	55–65	> 65
Observed frequency	10	18	28	18	12

Table 13.17

(a) Assuming that the random variable is modelled by a $N(50,15^2)$ distribution, calculate the expected frequencies for each of the five classes.

(b) Carry out a χ^2 goodness of fit analysis to test, at the 5% level, the hypothesis that H can be modelled as in part (a).   (OCR)

(a) If $H \sim N(50,15^2)$ then $Z \sim N(0,1)$ where $Z = \dfrac{H-50}{15}$. The values of Z corresponding to the class boundaries of 35, 45, 55 and 65 are -1, $-\frac{1}{3}$, $\frac{1}{3}$ and 1 respectively.

With a calculator the probabilities that Z lies in the five intervals separated by these class boundaries are 0.1587, 0.2108, 0.2611, 0.2108, 0.1587, correct to 4 decimal places.

To find the expected frequencies, multiply these probabilities by 86, the total number of conifers. This gives expected frequencies of 13.65, 18.13, 22.45, 18.13, 13.65.

(b) The null and alternative hypotheses are

$$H_0 : H \sim N(50,15^2); \qquad H_1 : H \text{ is not } N(50,15^2).$$

Table 13.18 gives the observed and expected frequencies for the five classes. The expected frequencies don't add up to exactly 86 because the probabilities in part (a) were rounded to 4 decimal places.

H	<35	35–45	45–55	55–65	>65	Total
f_o	10	18	28	18	12	86
f_e	13.65	18.13	22.45	18.13	13.65	86(.01)

Table 13.18

The calculation then proceeds just as in the discrete case. There are 5 classes and 1 constraint, that $\sum f_e = 86$, so $v = 4$. The value of X^2, calculated from Table 13.18, is 2.539... , and $P(\chi_4^2 \geq 2.539...) = 0.637...$, or 63.7%, much greater than the 5% significance level required to reject H_0. The null hypothesis, that H can be modelled by $N(50,15^2)$, is accepted.

To calculate expected frequencies for a normal distribution you need values of μ and σ^2. In the example above these values were given. If either or both of μ and σ^2 are not given, then unbiased estimates are used instead (see Section 10.1). (These estimates should be calculated from the grouped frequency table.) For each estimate used another constraint is added. For example, if both μ and σ^2 are estimated there will be 3 constraints in total.

Example 13.5.2

The lengths (in mm) of 50 leaves that had fallen from an oak were measured. The results are summarised in Table 13.19.

Length (mm)	30–39	40–49	50–59	60–69	70–79	80–89	90–99	100–109
Frequency	3	9	15	9	6	4	3	1

Table 13.19

By carrying out a goodness of fit test at the 5% significance level test whether the length of an oak leaf can be modelled by a normal distribution.

Notice that, in Table 13.19, the lengths are stated to the nearest millimetre. The class boundaries are therefore 29.5–39.5, 39.5–49.5,..., and the mid-class values are 34.5, 44.5, ...

Since the mean, μ, and the variance, σ^2, are not known, they must be estimated from the sample, basing the calculation on the mid-class values. Your calculator probably has a program for calculating $\bar{x}$ and s_{n-1}^2 from a data list, This gives $\bar{x} = 61.5$ and $s_{n-1} = 16.812$. So, if the lengths of leaves can be modelled by a normal distribution, the best guess is that it will be $N(61.5, 16.812^2)$.

So set the null and alternative hypotheses as follows.

H_0: The length, in mm, of oak leaves is distributed as $N(61.5, 16.812^2)$;

H_1: The length, in mm, of oak leaves is not distributed as $N(61.5, 16.812^2)$.

The next step is to calculate the expected frequencies. For this you need to know the probabilities that a random variable X lies in a particular interval, given the known values of μ and σ^2. Some calculators have a program which you can use to find these directly. Otherwise you standardise the class boundaries using the equation

$$Z = \frac{X - 61.5}{16.812}$$

and find the probabilities in the corresponding intervals for $N(0,1)$. To get the expected frequencies you then multiply these probabilities by 50. The results are shown in Table 13.20.

Class boundaries	< 29.5	29.5– 39.5	39.5– 49.5	49.5– 59.5	59.5– 69.5
Probability	0.0285	0.0668	0.1423	0.2150	0.2303
f_e	1.42	3.34	7.12	10.75	11.51

Class boundaries	69.5– 79.5	79.5– 89.5	89.5– 99.5	99.5– 109.5	> 109.5
Probability	0.1749	0.0943	0.0360	0.0098	0.0022
f_e	8.75	4.71	1.80	0.49	0.11

Table 13.20

You can see that it will be necessary to combine some classes, the first three and the last four. Table 13.21 shows the observed and expected frequencies with the classes combined.

Class boundaries	< 49.5	49.5– 59.5	59.5– 69.5	69.5– 79.5	> 79.5	Total
Probability	12	15	9	6	8	50
f_e	11.88	10.75	11.51	8.75	7.11	50

Table 13.21

There are 5 classes and 3 constraints ($\sum f_e = 50$, and μ and σ^2 estimated from the data), so $v = 5 - 3 = 2$. From Table 13.21, $X^2 = 3.204\ldots$, and

$$P\left(X^2 \geq 3.204\ldots\right) \approx P\left(\chi_2^2 \geq 3.204\ldots\right) = 0.2014\ldots \text{ , which is } 20.1\%.$$

This is greater than the significance level of 5%, so the null hypothesis is accepted. The length, in mm, of oak leaves is distributed as $N\left(61.5, 16.812^2\right)$.

Exercise 13B

1 It is thought that the random variable Y is distributed $N(50,100)$. One hundred observations of Y are made and result in the given frequency distribution.

Observations of Y	< 30	30–	40–	50–	60–	> 70
Frequency	3	14	30	35	14	4

It is decided to carry out a goodness of fit test. Some of the expected frequencies are given in the table below.

Value of Y	< 30	30–	40–	50–	60–	> 70
Expected frequency	2.28			34.13		

Complete this table and carry out a goodness of fit test at the 2.5% significance level.

2 Human intelligence is often measured by calculating the Intelligence Quotient (IQ) of an individual, where $IQ \sim N\left(100, 15^2\right)$. Three hundred new entrants to a school have IQ scores distributed according to the table.

IQ score	≤ 55	–70	–85	–100	–115	–130	–145	–160	> 160
Frequency	0	10	50	125	82	30	2	1	0

It is decided to test the fit of the model to these data. The expected frequencies (ignoring a continuity correction) are shown below.

IQ score	≤ 55	–70	–85	–100	–115	–130	–145	–160	> 160
Ex. frequency	0.41	6.42	40.77	102.40	102.40	40.77	6.42	0.40	0.01

(a) Show how the expected frequency for the fourth class was calculated.

(b) By carrying out a goodness of fit test at the 5% level, show that the model $IQ \sim N(100,15^2)$ is not suitable for these data.

(c) It is thought that a better model may be one in which μ is estimated from the data rather than taking $\mu = 100$. Obtain an estimate of μ from the data and test whether a normal distribution with this mean and $\sigma = 15$ gives a better fit to the data.

3   The table below shows the frequency distribution of the lifetimes of a sample of 50 light bulbs.

Lifetime (hours)	< 650	650–659	660–669	670–679	680–689	690–699
Frequency	0	1	3	3	7	15

Lifetime (hours)	700–709	710–719	720–729	730–739	> 739
Frequency	7	7	4	3	0

(a) Show that unbiased estimates of the mean and variance of the population from which this sample was drawn are 698.3 and 354.653 respectively.

It is thought that the lifetime of a light bulb can be modelled by a normal distribution.

(b) Calculate the expected frequencies for the model $N(698.3, 354.653)$ and carry out a goodness of fit test of this model at the 5% level.

4   The amounts of meat eaten per week in a sample of 100 families of four is shown in the table.

Amount of meat (kg)	0–	2–	4–	6–	8–10
Frequency	2	10	20	50	18

It is thought that the amount of meat eaten per week by such families can be modelled by the probability density function $f(w) = 0.0002w^3(10 - w)$, for $0 \le w \le 10$ and zero elsewhere.

(a) Show that the cumulative distribution function for this model is

$$F(w) = \begin{cases} 0 & \text{for } w < 0, \\ 0.000\,04w^4(12.5 - w) & \text{for } 0 \le w \le 10, \\ 1 & \text{for } w > 10. \end{cases}$$

(b) Show that the expected frequency for the class 8–10 is 26.27.

(c) Calculate the expected frequencies for the intervals in the table and conduct a goodness of fit test at the 10% level.

5   The heights of 50 male students aged 18 were recorded correct to the nearest centimetre. The results are summarised in the table.

Heights	156–160	161–165	166–170	171–175	176–180	181–185	186–190
Frequency	2	7	12	14	9	3	3

Find the sample mean and an unbiased estimate of population variance and use them to conduct a goodness of fit test at the 5% level, to see whether height is distributed normally.

6 The durations (in minutes) of 100 phone calls were recorded, resulting in the frequency distribution below.

Duration of calls	0–	1–	2–	3–	4–	5–	10–	20
Frequency	7	18	34	25	13	2	1	0

Are the data normally distributed? Conduct a goodness of fit test at 0.1% level of significance.

13.6 Contingency tables

This section turns to a different kind of significance test which makes use of the χ^2 distribution. The question is whether two attributes of a population tend to be associated, that is to occur together, or whether they occur independently of each other.

Table 13.22 gives some data relating to the income level and the method of getting to work for a random sample of people. The income level is described as 'small' (less than £15 000), 'average' (£15 000 to £25 000), or 'large' (greater than £25 000) and the way of getting to work as 'car' (travelling by car either as driver or passenger), 'public' (using public transport) or 'self' (either cycling or walking). Thus the two attributes are income level and method of transport. A table like this, in which the total frequency has been divided into rows and columns according to the values taken by two attributes, is called a **contingency table**.

		Method of transport			
		Car	Public	Self	Total
	Small	58	21	36	115
Income	Average	199	49	64	312
level	Large	205	32	29	266
	Total	462	102	129	693

Table 13.22

The row and column totals allow you to estimate probabilities of the values for the different attributes. For example, $P(\text{small}) = \frac{115}{693}$ and $P(\text{public}) = \frac{102}{693}$.

Are the variables 'income level' and 'method of transport' independent of each other? This question can be tested by calculating the frequencies which you would expect if the variables *are* independent and then carrying out a χ^2 test. The null and alternative hypotheses are

H_0: the variables 'income level' and 'method of transport' are independent;
H_1: the variables 'income level' and 'method of transport' are not independent.

Suppose you want to find the expected frequency for the group of people with small income who travel by public transport. In order to do this you need an estimate of $P(\text{small and public})$. If H_0 is true and the variables are independent, then

$$P(\text{small and public}) = P(\text{small}) \times P(\text{public}) = \frac{115}{693} \times \frac{102}{693}.$$

Since the total sample size is 693, the corresponding expected frequency is equal to $693 \times \frac{115}{693} \times \frac{102}{693} = 16.93$, correct to 2 decimal places.

The other expected frequencies can be found in a similar fashion. For example,

$$P(\text{large and self}) = P(\text{large}) \times P(\text{self}) = \frac{226}{693} \times \frac{129}{693},$$

so the expected frequency is $693 \times \frac{226}{693} \times \frac{129}{693} = 49.52$, correct to 2 decimal places.

There is a quicker way of calculating the expected frequencies. Look at $693 \times \frac{115}{693} \times \frac{102}{693}$, which is the expression for the expected frequency for people with small income travelling by public transport. Two of the 693s cancel to give $\frac{115 \times 102}{693}$. The numerator of this fraction is the product of the row total for small income and the column total for public transport, while the denominator is the total sample size, usually called the 'grand total'. This is an example of a general rule for finding expected frequencies.

> Expected frequencies for a contingency table are given by
>
> $$\text{expected frequency} = \frac{\text{row total} \times \text{column total}}{\text{grand total}}.$$

Table 13.23 shows the calculation of the expected frequencies using this formula.

		Method of transport			
		Car	Public	Self	Total
	Small	$\frac{115 \times 462}{693} = 76.67$	$\frac{115 \times 102}{693} = 16.93$	$\frac{115 \times 129}{693} = 21.41$	115
Income	Average	$\frac{312 \times 462}{693} = 208.00$	$\frac{312 \times 102}{693} = 45.92$	$\frac{312 \times 129}{693} = 58.08$	312
level	Large	$\frac{266 \times 462}{693} = 177.33$	$\frac{266 \times 102}{693} = 39.15$	$\frac{266 \times 129}{693} = 49.52$	266
	Total	462	102	129	693

Table 13.23

A value of X^2 can now be calculated from the observed and expected frequencies as shown in Table 13.24. The total of the expected frequencies is not equal to 693 exactly because of rounding errors.

Income/transport	S/C	S/P	S/S	A/C	A/P	A/S	L/C	L/P	L/S
f_o	58	21	36	199	49	64	205	32	29
f_e	76.67	16.93	21.41	208.00	45.92	58.08	177.33	39.15	49.52

Table 13.24

From this you can work out that $X^2 = 30.78$

You now need to know the number of degrees of freedom of the χ^2 distribution which will approximate the distribution of X^2. As for goodness of fit tests this is calculated from

ν = number of classes − number of constraints.

There are 9 classes, but how many constraints are there? Look at Table 13.23 again. In this table all three column totals, all three row totals and the grand total were used to calculate the expected frequencies. However, the calculation could have been done with fewer pieces of information as shown in Table 13.25. The missing row and column totals in this table can be found by subtraction and then the expected frequencies calculated as before (check that you agree with this statement). So this table shows that there are 5 constraints, because 5 pieces of information from the observed contingency table are needed in order to calculate the expected frequencies. Thus $\nu = 9 - 5 = 4$.

		Method of transport			
		Car	Public	Self	Total
Income level	Small				115
	Average				312
	Large				
	Total	462	102		693

Table 13.25

So now calculate $P(X^2 \geq 30.78) \approx P(\chi_4^2 \geq 30.78) = 3.4 \times 10^{-6}$. Thus the null hypothesis that the two variables are independent can be rejected at any reasonable significance level: there is very strong evidence that income size and method of transport are related.

Having found that income level and method of transport are not independent it is interesting to see if there is any obvious reason for this. If you look at Table 13.24 and work out X^2 using the formula

$$\sum \frac{(f_o - f_e)^2}{f_e},$$ *you will see that large contributions to the value of X^2 are made by the first, third, seventh*

and ninth classes. If you then compare the observed and expected frequencies for these classes you will see that among people with small incomes fewer than expected travel to work by car and more than expected walk or cycle. Among people with large incomes the opposite is true.

13.7 Degrees of freedom of a contingency table

The method used in the previous section to find the number of degrees of freedom of a contingency table can be generalised. If there are h rows and k columns, then the number of classes is hk. Provided you are given all but one of the column totals, all but one of the row totals and the grand total, you can find all the expected frequencies, using subtraction where appropriate. So the number of constraints is equal to $(h-1) + (k-1) + 1 = h + k - 1$. Thus

$\nu =$ number of classes $-$ number of constraints

$= hk - (h + k - 1) = hk - h - k + 1$

$= (h-1)(k-1).$

> The number of degrees of freedom of a contingency table is given by
>
> $$\nu = (\text{number of rows} - 1) \times (\text{number of columns} - 1).$$

Applying this rule to the example in Section 13.6, where $h = k = 3$, gives $\nu = (3-1) \times (3-1) = 4$, as before.

13.8 Combining rows and columns

As for other goodness of fit tests, each expected frequency must be at least 5 in order to carry out a χ^2 test for independence of the two attributes. If necessary, rows or columns in the contingency table can be combined in order to meet this condition.

Note that you should only combine whole rows or columns, not individual entries.

Example 13.8.1

A university sociology department believes that students with a good grade in A level General Studies tend to do well on sociology degree courses. To check this it collected information on a random sample of 100 students who had just graduated and had also taken General Studies at A level. The students' performance in General Studies was divided into two categories, those with grade A or B and 'others'. Their degree classes were recorded as Class I, Class II, Class III and Fail. The data are in Table 13.26.

	Class I	Class II	Class III	Fail	Total
Grade A or B	11	22	6	1	40
Others	4	28	24	4	60
Total	15	50	30	5	100

Table 13.26

Use these data to test, at the 1% significance level, the hypothesis that degree class is independent of General Studies A level performance. State your conclusion clearly.   (OCR, adapted)

Take

H_0: degree class is independent of General Studies A level performance;
H_1: degree class is not independent of General Studies A level performance.

Table 13.27 shows the calculation of the expected frequencies, assuming independence.

	Class I	Class II	Class III	Fail	Total
Grade A or B	$\frac{40\times15}{100}=6$	$\frac{40\times50}{100}=20$	$\frac{40\times30}{100}=12$	$\frac{40\times5}{100}=2$	40
Others	$\frac{60\times15}{100}=9$	$\frac{60\times50}{100}=30$	$\frac{60\times30}{100}=18$	$\frac{60\times5}{100}=3$	60
Total	15	50	30	5	100

Table 13.27

Minimum expected frequencies of 5 can be achieved by combining the columns for Class III and Fail, that is, in each row, adding the frequencies for Class III and Fail.

This combination is logical because it groups 'weak' students together. It would not be sensible, for example, to combine Class I with Fail. Note also that in this example it would be pointless to combine rows since there would then only be one category for General Studies grade. In general, the combination of rows and columns should always be done on the basis of common sense.

Table 13.28 gives the observed and expected frequencies after combination. The values in brackets are the expected frequencies.

	Class I	Class II	Class III and Fail
Grade A or B	11(6)	22(20)	7(14)
Others	4(9)	28(30)	28(21)

Table 13.28

This gives the value $X^2 = 13.111\ldots$.

The number of degrees of freedom is calculated after any combination of rows or columns has been made in the contingency table. So applying the formula in Section 13.7 to Table 13.28,

$$v = (\text{number of rows} - 1) \times (\text{number of columns} - 1)$$
$$= (2-1)(3-1) = 2.$$

So $P(X^2 \geq 13.111\ldots) \approx P(\chi_2^2 \geq 13.111\ldots) = 0.001\,42\ldots$, which is 0.14%. This is less than the significance level of 1%, so 13.111... is in the rejection region.

The null hypothesis that degree class is independent of General Studies A level performance is rejected.

Inspection of Table 13.28 suggests that Class I degree results tend to be associated with grade A or B in General Studies while Class III degrees tend to be associated with lower grades in General Studies.

13.9* Justification for the χ^2 test procedure

All the applications in this chapter have been based on the link between the test statistic $\sum \dfrac{(f_o - f_e)^2}{f_e}$ and χ_v^2 for some value of v. The full theory supporting this can't be given in this book, but here is a proof for the special case $v = 1$. You may omit it if you wish.

The definition of χ^2 is that, if Z_1, Z_2, $\ldots$, Z_v are v independent variables, each with a N(0,1) distribution, then the sum of their squares, that is $\sum Z_i^2$, has a χ_v^2 distribution. In this chapter you have used the result that the statistic $\sum \dfrac{(f_o - f_e)^2}{f_e}$ has a distribution which is approximately χ_v^2 where there are v degrees of freedom. It is fairly simple to show that this is true for the particular case in which there is one degree of freedom. In this case $\sum \dfrac{(f_o - f_e)^2}{f_e} \sim \chi_1^2$, so $\sum \dfrac{(f_o - f_e)^2}{f_e} = Z^2$.

Consider the situation in which the results fall into one of two classes. Suppose the null hypothesis is that the values are divided between the two classes in the ratio $p:(1-p)$. Denote the observed frequencies by O_1 and O_2, then the total frequency, n, is given by $n = O_1 + O_2$. The corresponding expected frequencies are np and $n(1-p)$. Then

$$\sum \frac{(f_e - f_o)^2}{f_e} = \frac{(O_1 - np)^2}{np} + \frac{(O_2 - n(1-p))^2}{n(1-p)}$$

$$= \frac{(O_1 - np)^2}{np} + \frac{(n - O_1 - n + np)^2}{n(1-p)} \qquad \text{(since } O_2 = n - O_1)$$

$$= \frac{(O_1 - np)^2}{np} + \frac{(np - O_1)^2}{n(1-p)}$$

$$= \frac{(O_1 - np)^2}{n} \left(\frac{1}{p} + \frac{1}{1-p} \right)$$

$$= \frac{(O_1 - np)^2}{n} \left(\frac{1-p+p}{p(1-p)} \right)$$

$$= \frac{(O_1 - np)^2}{np(1-p)}.$$

Now O_1 is distributed as $B(n, p)$ since there are n trials and the probability for a result in the first class is p. If n is such that np and $n(1-p)$ are not too small, then this binomial distribution can be approximated by $N(np, np(1-p))$. Thus $O_1 \sim N(np, np(1-p))$ approximately. Standardising this distribution gives

$$Z = \frac{O_1 - np}{\sqrt{np(1-p)}} \sim N(0,1).$$

Squaring gives $Z^2 = \dfrac{(O_1 - np)^2}{np(1-p)}$. It was shown above that $\sum \dfrac{(f_o - f_e)^2}{f_e} = \dfrac{(O_1 - np)^2}{np(1-p)}$ so, in this case,

$\sum \dfrac{(f_o - f_e)^2}{f_e} = Z^2$, approximately.

This result is consistent with the rule for calculating the number of degrees of freedom since there are 2 classes and 1 constraint; that is, that the total observed frequency n is used in the calculation of the expected frequencies.

Exercise 13C

1   The grades achieved at GCSE in maths and English of 500 students are summarised in the table.

		Maths grade		
		A or A*	B	C or worse
English	A or A*	15	30	15
grade	B	33	105	102
	C or worse	27	65	108

Conduct a χ^2 test at the 5% significance level to test the hypothesis that the grades achieved in maths and English are independent.

2   One hundred and thirty people in London were asked to identify their favourite brand of soap from among Brand X, Brand Y and Brand Z. Seventy people in Manchester were asked the same question. The results of the surveys are shown in the table.

	Brand X	Brand Y	Brand Z
London	35	70	25
Manchester	15	35	20

Is there any association between the city and the brand of soap preferred? Use a 5% level of significance.

3   The milk yield from two breeds of cows is classified as 'High', 'Medium' or 'Low'. The yields of 100 cows were classified as in the table.

	Yield		
	High	Medium	Low
Breed A	30	20	18
Breed B	18	10	4

Use these data to test, at the 2.5% level of significance, whether breed of cow and milk yield are independent.

4   Six hundred individuals were classified according to their hair colour and eye colour.

		Eye colour		
		Brown	Green/Grey	Blue
	Black	95	56	44
Hair	Brown	78	64	90
colour	Fair	37	35	61
	Ginger	10	10	20

Conduct a χ^2 test at the 0.1% significance level to test the hypothesis that hair colour and eye colour are independent.

5   The table below gives some data which were obtained from a random sample of men in the course of a study into heart disease. Each man was classified according to his income level and his level of physical activity. The amount of exercise is described as 'Low', 'Medium' or 'High' and the income level as 'Small', 'Average' or 'Large'.

		Level of physical activity		
		Low	Medium	High
Income	Small	143	56	108
level	Average	134	130	188
	Large	53	82	106

Use these data to test, at the 1% significance level, the hypothesis that level of physical activity and income level are independent.

6 The grade achieved at A-level mathematics is recorded for a random sample of 100 students studying at one of three sixth-form colleges in a city. The data are summarised in the table.

		Grade						
		A	B	C	D	E	N	U
College	X	12	34	23	7	5	4	1
	Y	4	45	12	5	3	4	2
	Z	27	23	16	15	5	2	1

It is proposed to test whether there is any association between the college attended and the grade obtained.

(a) Calculate the expected frequencies for a χ^2 test, assuming no association between grade and college.

(b) Explain briefly why some combining of rows or columns should be carried out.

(c) Carry out the test, combining suitable columns and using a 1% significance level.

7 In an investigation into a possible association between hair colour and height, a random sample of 200 adult men was taken, and the data shown in the table were obtained.

		Hair colour			
		Dark	Fair	Red	Total
Height	Less than 165 cm	16	7	11	34
	165 cm to 180 cm	46	39	9	94
	More than 180 cm	33	35	4	72
	Total	95	81	24	200

It is proposed to carry out a χ^2 test for independence between hair colour and height.

(a) Calculate the expected frequencies.

(b) Explain briefly why some combining of rows or columns should be carried out.

(c) Carry out the test, combining suitable rows and using a 5% significance level.   (OCR)

Review exercise

1   X is a random variable with the distribution $X \sim B(12, 0.42)$.

   (a) Anne uses the binomial distribution to calculate the probability that $X < 4$ and gives 4 significant figures in her answer. What answer should she get?

   (b) Ben uses a normal distribution to calculate an approximation for the probability that $X < 4$ and gives 4 significant figures in his answer. What answer should he get?   (OCR, adapted)

2   The time T hours taken to repair a piece of equipment has a probability density function which may be modelled by

$$f(t) = \begin{cases} \dfrac{24}{7t^4} & \text{for } 1 \le t \le 2, \\ 0 & \text{otherwise.} \end{cases}$$

   (a) Find $E(T)$ and $\text{Var}(T)$.

   (b) $\overline{T}$ denotes the mean of the times for 30 randomly chosen repairs. Assuming that the central limit theorem holds, estimate $P(\overline{T} < 1.2)$.

   State, giving a reason, whether your answer has little error or considerable error.

3   A machine is set to produce ball-bearings with mean diameter 1.2 cm. Each day a random sample of 50 ball-bearings is selected and the diameters accurately measured. If the sample mean diameter lies outside the range 1.18 cm to 1.22 cm then it will be taken as evidence that the mean diameter of the ball-bearings produced is not 1.2 cm. The machine will then be stopped and adjustments made to it. Assuming that the diameters have standard deviation 0.075 cm, find the probability that

   (a) the machine is stopped unnecessarily,

   (b) the machine is not stopped when the mean diameter of the ball-bearings produced is 1.15 cm.

4   The number of night calls to a fire station serving a small town can be modelled by a Poisson distribution with mean 2.7 calls per night.

   (a) State the expectation and variance of the mean number of night calls over a period of n nights.

   (b) Estimate the probability that during a given year of 365 days the total number of night calls will exceed 1050.

5   Metal struts used in a building are specified to have a mean length of 2.855 m. The lengths have a normal distribution with standard deviation 0.0352 m. A batch of 15 struts is sent to a building site and the lengths are measured. The sample mean length is 2.841 m.

   A test is to be carried out, at the 5% significance level, to decide whether the batch is from the specified population.

   (a) Stating your hypotheses, find the rejection region in terms of Z.

   (b) State the conclusion of the test.

6 A supermarket buys a large batch of plastic bags from a manufacturer to be used in the store. In previous batches 7% of the bags were defective. A quality control manager wishes to test whether the batch has a higher defective rate than 7%, in which case the batch will be returned to the manufacturer. He examines 125 randomly selected bags and finds that 14 are defective. Using a distributional approximation, carry out the manager's test at the 3% significance level and state whether he should return the batch.

7 An employee is accused by his employer of being late for work too often. The employee claims that, on average, he is late on no more than one day in ten. The employer finds that, over a random sample of 20 days, the employee is late on r days. The employer carries out a significance test, at the 5% level, to decide whether, on average, the employee is late on more than one day in ten.

(a) State suitable null and alternative hypotheses for the test.

(b) Find the set of values of r for which the null hypothesis would be rejected, and state the conclusion of the test in the case $r = 4$.

(c) Given that, in fact, the probability that the employee is late for work on a randomly chosen day is 0.2, find the probability of making a Type II error in the test.   (OCR)

8 The time, T, in minutes that a person takes to drive to work is uniformly distributed over the interval 30 to 40. The distance of the journey is 15 km. Let V denote the average speed of the journey in km per hour.

(a) Write down the relationship between T and V.

(b) Find the probability density function of V.

(c) Find (i) the median (ii) the mean of V.

9 Small packets of nails are advertised as having average weight 500 g, and large packets as having average weight 1000 g. Assume that the packet weights are distributed normally with means as advertised, and standard deviations of 10 g for a small packet and 15 g for a large packet. Giving your answers correct to 3 decimal places,

(a) find the probability that two randomly chosen small packets have a total weight between 990 g and 1020 g,

(b) find the probability that the weight of one randomly chosen large packet exceeds the total weight of two randomly chosen small packets by at least 25 g,

(c) find the probability that one half of the weight of one randomly chosen large packet exceeds the weight of one randomly chosen small packet by at least 12.5 g.   (OCR)

10 A slimming regime, consisting of a mixture of diet and exercise, is claimed by the designer to decrease the weight of people using the regime by an average of 4 kg over a three-month period. To support the claim, the designer measures the weight losses of 16 randomly chosen people who used the regime for three months. The results, w kg, are summarised by $\sum w = 46.44$ and $\sum w^2 = 180.5$.

(a) Test, at the 10% significance level, the null hypothesis that the population mean weight loss is 4 kg against the alternative hypothesis that it is not 4 kg. Weight loss may be assumed to be distributed normally.

(b) Find the set of values of the population mean weight loss for which the null hypothesis would be rejected in a two-tail test at the 5% significance level using the above data.

11 Ten students were asked to perform a simple task and the times, in seconds, that they took to complete the task were recorded. The next day, the students were asked to perform the same task and the times taken on the second attempt were recorded. A researcher was interested to know whether the practice in the first attempt led to a reduction in time at the second attempt. The times are given in the following table.

Student	1	2	3	4	5	6	7	8	9	10
First attempt	16.3	54.2	63.2	42.0	38.7	48.3	11.2	41.3	52.1	44.9
Second attempt	13.2	57.1	50.1	40.0	36.0	30.2	9.6	44.0	40.3	23.8

(a) Stating your assumptions, use a suitable test, at the 5% significance level, to advise the researcher as to whether there is a reduction in time at the second attempt.

(b) Calculate a 95% confidence interval for the population mean reduction in time for the two attempts.   (OCR, adapted)

12 In a test of car components in which the failure rate is thought to be constant, the numbers of failures in a series of 100-mile intervals are recorded in the following table.

Number of failures	0	1	2	3	4	5	6	7
Frequency	25	30	26	18	9	5	6	1

Calculate the mean number of failures per interval and the expected distribution of failures for the 120 intervals that would be given by a Poisson distribution with this mean.

A good fit between the sets of observed and expected frequencies is taken as evidence of the constancy of the failure rate. Use a χ^2 test to test the goodness of fit.   (OCR)

13 In 1988 the number of new cases of insulin-dependent diabetes in children under the age of 15 years was 1495. The table below breaks down this figure according to age and sex.

	Age (years)			
	0–4	5–9	10–14	Total
Boys	205	248	328	781
Girls	182	251	281	714
Total	387	499	609	1495

Perform a suitable test, at the 5% significance level, to determine whether age and sex are independent factors.   (OCR)

Examination questions

1 The weights of male nurses in a hospital are known to be normally distributed with mean $\mu = 72$ kg and standard deviation $\sigma = 7.5$ kg. The hospital has a lift (elevator) with a maximum recommended load of 450 kg. Six male nurses enter the lift. Calculate the probability p that their combined weight exceeds the maximum recommended load.   (© IBO 2002)

2 The following is a random sample of 16 measurements of the density of aluminium. Assume that the measurements are normally distributed.

2.704	2.709	2.711	2.706
2.708	2.705	2.709	2.701
2.705	2.707	2.710	2.700
2.703	2.699	2.702	2.701

Construct a 95% confidence interval for the density of aluminium, showing all steps clearly.

(© IBO 2002)

3 A sociologist wants to know whether the percentage of sons taking up the profession of their father is the same in every profession. She decides to investigate the situation in each of four professions. She obtained the following data.

63 out of 136 sons of male medical doctors became doctors

42 out of 118 sons of male engineers became engineers

35 out of 96 sons of male lawyers became lawyers

68 out of 150 sons of male businessmen became businessmen

At the 5% level of significance what should her conclusion be?   (© IBO 2002)

4 *Give all numerical answers to this question correct to two decimal places.*

A radar records the speed, v kilometres per hour, of cars on a road. The speed of these cars is normally distributed. The results for 1000 cars are recorded in the following table.

Speed	Number of cars
$40 \leq v < 50$	9
$50 \leq v < 60$	35
$60 \leq v < 70$	93
$70 \leq v < 80$	139
$80 \leq v < 90$	261
$90 \leq v < 100$	295
$100 \leq v < 110$	131
$110 \leq v < 120$	26
$120 \leq v < 130$	11

(a) For the cars on the road, calculate

 (i) an unbiased estimate of the mean speed;

 (ii) an unbiased estimate of the variance of the speed.

(b) For the cars on the road, calculate

 (i) a 95% confidence interval for the mean speed;

 (ii) a 90% confidence interval for the mean speed.

(c) Explain why one of the intervals found in part (b) is a subset of the other.   (© IBO 2003)

5 Eggs at a farm are sold in boxes of six. Each egg is either brown or white. The owner believes that the number of brown eggs in a box can be modelled by a binomial distribution. He examines 100 boxes and obtains the following data.

Number of brown eggs in a box	Frequency
0	10
1	29
2	31
3	18
4	8
5	3
6	1

(a) (i) Calculate the mean number of brown eggs in a box.

(ii) Hence estimate p, the probability that a randomly chosen egg is brown.

(b) By calculating an appropriate statistic, test, at the 2.5% significance level, whether or not the binomial distribution gives a good fit to these data. (© IBO 2003)

6 The random variable Z has probability density function $f(z) = z - \frac{1}{4}z^3$ for $z \in [0,2]$ and 0 elsewhere. A physicist assumes that the lifetime of a certain particle can be modelled by this random variable. The interval $[0, 2]$ is divided into the following equal intervals:

$$I_1 = [0, 0.4[\qquad I_2 = [0.4, 0.8[\qquad I_3 = [0.8, 1.2[\qquad I_4 = [1.2, 1.6[\qquad I_5 = [1.6, 2]$$

The physicist carried out 40 experiments and recorded the number of times the value of Z lay in each of the intervals I_k where $k = 1,2,3,4,5$ as shown in the following table:

I_1	I_2	I_3	I_4	I_5
2	12	9	8	9

(a) Assuming that the physicist's assumption is correct, for each value of k find $p_k = P(Z \in I_k)$.

(b) At the 5% significance level can his assumption be accepted? (© IBO 2004)

7 Let X and Y be two independent variables with $E(X) = 5$, $\text{Var}(X) = 3$, $E(Y) = 4$, $\text{Var}(Y) = 2$. Find

(a) $E(2X)$, (b) $\text{Var}(2X)$, (c) $E(3X - 2Y)$, (d) $\text{Var}(3X - 2Y)$.

(© IBO 2005)

8 A machine shop manufactures steel rods for use in a car production plant. The lengths in metres for a sample of 8 rods are given below.

0.999, 1.001, 1.005, 1.011, 1.005, 1.001, 0.998, 1.004

Previous observations have shown that the machine settings gave rods with lengths that are normally distributed with standard deviation 0.0028 m.

Stating the type of test used, determine at the 1% significance level if the mean length of the rods produced is 1.005 m. (© IBO 2005)

Appendix: some supporting mathematics

In this book it has been possible to prove some results in probability using mathematics you have already met in the Higher Level books. Other results, such as the central limit theorem or the equation for t-probability, require far more advanced mathematics and must at this stage be taken on trust. This appendix deals with a few mathematical techniques which come between these extremes. They are not an essential part of the course, but they are included so that you can, if you wish, fill in some gaps in the treatment of certain topics.

Most of the material in this appendix takes the form of questions for you to work for yourself. For ease of reference questions are numbered as **Qx.y**, where the number x indicates the section in which it appears, and y tells you the position of the question within the section.

A.1 Mean and variance

This section deals with discrete probability distributions for which the sample space S is $\mathbb{N}$ or a subset of $\mathbb{N}$, such as $\mathbb{Z}^+$ or the finite set $\{0, 1, 2, \ldots, n\}$. If p_x denotes the probability $P(X = x)$, then the standard definitions can be written as

$$\mu = \sum_{x \in S} x p_x \quad \text{and} \quad \sigma^2 = \sum_{x \in S} (x - \mu)^2 p_x, \quad \text{where} \quad \sum_{x \in S} p_x = 1.$$

There are two other ways of finding the variance. You have often used the formula in Q1.1, but Q1.2 may be new to you.

Q1.1 Prove that $\sigma^2 = \sum_{x \in S} x^2 p_x - \mu^2$.

Q1.2 Prove that $\sigma^2 = \sum_{x \in S} x(x - 1) p_x + \mu - \mu^2$.

A.2 The negative binomial series

One way of writing the binomial expansion of $(1 + x)^n$, where $n \in \mathbb{Z}^+$, is

$$(1 + x)^n = 1 + \frac{n}{1!} x + \frac{n(n - 1)}{2!} x^2 + \frac{n(n - 1)(n - 2)}{3!} x^3 + \ldots + x^n.$$

If you substitute $x = -q$ and $n = -1$ in this equation, the left side becomes $(1 - q)^{-1}$ but the series on the right becomes an infinite series.

Q2.1 Write down the right side with these substitutions, simplifying the coefficients as much as possible. Where have you seen this before?

Q2.2 Show that, if you carry out the the same process with $n = -2$ and $n = -3$, you get

$$(1 - q)^{-2} = 1 + 2q + 3q^2 + \ldots = \sum_{x=0}^{\infty} x q^{x-1} \quad \text{and} \quad (1 - q)^{-3} = 1 + 3q + 6q^2 + \ldots = \sum_{x=0}^{\infty} x(x - 1) q^{x-2}.$$

Q2.3 Take the expressions on both sides of Q2.1 and differentiate them once and twice with respect to q. (It can be proved that this is valid if $-1 < q < 1$.) Compare your results with those in Q2.2.

Q2.4 Use a calculator to display with the same axes the graphs of $y = (1-x)^{-2}$ and $y = 1 + 2x + 3x^2 + 4x^3 + 5x^4 + 6x^5$. Use a window of $-2 \le x \le 2, -5 \le y \le 10$. For what interval of values of x do the two graphs appear to coincide?

Q2.5 Repeat Q2.4 with the graphs of $y = (1-x)^{-3}$ and $y = 1 + 3x + 6x^2 + 10x^3 + 15x^4 + 21x^5$.

Q2.4 and Q2.5 suggest (as is true) that the equations in Q2.2 are valid if $-1 < q < 1$.

Q2.6 Assuming this, and using the expression for σ^2 in Q1.2, prove that, for the geometric distribution with $P(X = x) = pq^{x-1}$ for $x \in \mathbb{Z}^+$, $\mu = \dfrac{1}{p}$ and $\sigma^2 = \dfrac{q}{p^2}$. (See Section 2.2.)

If you write a series like those in Q2.2 with $n = -r$, you get

$$(1-q)^{-r} = 1 + \frac{r}{1!}q + \frac{r(r+1)}{2!}q^2 + \frac{r(r+1)(r+2)}{3!}q^3 + \dots .$$

Q2.7 Show that the coefficient of q^s can be written as $\begin{pmatrix} r+s-1 \\ r-1 \end{pmatrix}$. Hence the coefficient of q^{x-r} is $\begin{pmatrix} x-1 \\ r-1 \end{pmatrix}$.

This explains the name 'negative binomial' for the probability distribution $\mathrm{NB}(r, p)$ defined by

$$P(X = x) = \begin{pmatrix} x-1 \\ r-1 \end{pmatrix} q^{x-r} p^r. \text{ (See Section 12.2(iv)).}$$

A.3 The exponential series

Infinite series like those in Q2.2 are called power series, because each term is a multiple of a power of q. Another important power series is

$$E(m) = 1 + \frac{m}{1!} + \frac{m^2}{2!} + \frac{m^3}{3!} + \dots = \sum_{x=0}^{\infty} \frac{m^x}{x!}.$$

This series converges to a limit for all values of m.

Q3.1 Assuming that this series can be differentiated in the same way as in Q2.3, prove that

$$E'(m) = E(m).$$

Q3.2 Use Q3.1 and the fact that $E(0) = 1$ to prove that $E(m) = e^m$. (See Higher Level Book 2 Section 23.3.)

Q3.3 Show that, for Poisson probability $\mathrm{Po}(m)$, with $p_x = e^{-m} \dfrac{m^x}{x!}$, $\sum_{x=0}^{\infty} p_x = 1$.

Q3.4 Using Q1.2, prove that, for Poisson probability $\mathrm{Po}(m)$, $\mu = m$ and $\sigma^2 = m$.

A.4 Normal probability density

The equation for standardised normal probability density $N(0,1)$ is $\phi(x) = \dfrac{1}{\sqrt{2\pi}}e^{-\frac{1}{2}x^2}$.

Q4.1 Use the chain rule to find $\dfrac{d}{dx}e^{-\frac{1}{2}x^2}$.

Q4.2 Find the coordinates of the points of inflexion on the graph of $y = \phi(x)$.

Q4.3 By writing $x^2 e^{-\frac{1}{2}x^2}$ as $(-x) \times \left(-xe^{-\frac{1}{2}x^2}\right)$ and using integration by parts, prove that

$$\int_{-v}^{v} x^2 e^{-\frac{1}{2}x^2}\, dx = -2ve^{-\frac{1}{2}v^2} + \int_{-v}^{v} e^{-\frac{1}{2}x^2}\, dx.$$

Q4.4 Prove that $\displaystyle\int_{-\infty}^{\infty} x^2 \phi(x)\, dx = \int_{-\infty}^{\infty} \phi(x)\, dx$. Hence show that the variance of this probability

distribution is equal to 1, and give a geometrical interpretation of this in terms of the graph of $y = \phi(x)$.

Q4.5 Use a method similar to that in Q4.3 and Q4.4 to show that $\displaystyle\int_{-\infty}^{\infty} x^4 \phi(x)\,dx = 3\int_{-\infty}^{\infty} x^2 \phi(x)\,dx$.

Q4.6 By writing $\mathrm{Var}\!\left(Z^2\right)$ as $E\!\left(Z^4\right) - \left(E\!\left(Z^2\right)\right)^2$, prove that $\mathrm{Var}\!\left(Z^2\right) = 2$.

In Section 13.9 it is stated that, if $Z_1, Z_2, \ldots, Z_v$ are v independent variables, each with a $N(0,1)$

distribution, then $\displaystyle\sum_{i=1}^{v} Z_i^2$ has a χ_v^2 distribution.

Q4.7 Use the answer to Q4.6 to show that $E\!\left(\chi_1^2\right) = 1$ and $\mathrm{Var}\!\left(\chi_1^2\right) = 2$; and, more generally, that $E\!\left(\chi_v^2\right) = v$ and $\mathrm{Var}\!\left(\chi_v^2\right) = 2v$, as stated in Section 13.2.

A.5 Sampling without replacement

Q5.1 By expanding both sides of the identity $(1+y)^M (1+y)^{N-M} \equiv (1+y)^N$ and equating the

coefficients of y^n, prove that $\displaystyle\sum_{x=0}^{n} \binom{M}{x}\binom{N-M}{n-x} = \binom{N}{n}$.

Q5.2 For the hypergeometric probability distribution $\mathrm{Hyp}(n, M, N)$, with

$$p_x = \frac{\dbinom{M}{x}\dbinom{N-M}{n-x}}{\dbinom{N}{n}} \qquad \text{for } x = 0, 1, 2, 3, \ldots, n,$$

show that $\displaystyle\sum_{x=0}^{n} p_x = 1$. (See Section 12.3.)

Q5.3 Write an expression for the mean μ of this probability distribution. By using the identity $x\begin{pmatrix} M \\ x \end{pmatrix} \equiv M\begin{pmatrix} M-1 \\ x-1 \end{pmatrix}$, show that this mean can be written as $\mu = \dfrac{M}{\begin{pmatrix} N \\ n \end{pmatrix}} \sum\limits_{x=1}^{n}\begin{pmatrix} M-1 \\ x-1 \end{pmatrix}\begin{pmatrix} N-M \\ n-x \end{pmatrix}$.

Q5.4 Use a method like that in Q5.1 to prove that the sum in Q5.3 is equal to $\begin{pmatrix} N-1 \\ n-1 \end{pmatrix}$.

Hence show that $\mu = np$, where $p = \dfrac{M}{N}$.

Q5.5 By methods similar to those in Q5.2 to Q5.4, prove that

$$\sum_{x=0}^{n} x(x-1)p_x = \frac{M(M-1)}{\begin{pmatrix} N \\ n \end{pmatrix}} \sum_{x=2}^{n}\begin{pmatrix} M-2 \\ x-2 \end{pmatrix}\begin{pmatrix} N-M \\ n-x \end{pmatrix} = \frac{M(M-1)}{\begin{pmatrix} N \\ n \end{pmatrix}}\begin{pmatrix} N-2 \\ n-2 \end{pmatrix} = \frac{M(M-1)}{N(N-1)} \times n(n-1).$$

Q5.6 Hence, using the form for the variance in Q1.2, prove that $\sigma^2 = np(1-p)\left(\dfrac{N-n}{N-1}\right)$.

A.6 Exponential probability

The equation for exponential probability density is

$$f(x) = \begin{cases} me^{-mx} & \text{for } x > 0, \\ 0 & \text{otherwise.} \end{cases}$$

The mean and variance are therefore given by

$$\mu = \int_0^\infty mxe^{-mx}\,dx \quad \text{and} \quad \sigma^2 = \int_0^\infty mx^2 e^{-mx}\,dx - \mu^2.$$

Q6.1 Use integration by parts to find an expression for $\displaystyle\int_0^v mxe^{-mx}\,dx$ in terms of v.

Q6.2 Prove that $\displaystyle\int_0^v mx^2 e^{-mx}\,dx = -v^2 e^{-mv} + 2\int_0^v xe^{-mx}\,dx$.

In Higher Level Book 2 Review exercise 4 Question 20 you were asked to prove that $\lim\limits_{v\to\infty} ve^{-mv} = 0$ and $\lim\limits_{v\to\infty} v^2 e^{-mv} = 0$.

Q6.3 Use these limits together with Q6.1 and Q6.2 to prove that, for exponential probability $\text{Exp}(m)$,

$$\mu = \frac{1}{m} \text{ and } \sigma^2 = \frac{1}{m^2}. \text{ (See Section 2.4.)}$$

Answers

1 Cumulative distribution functions

Exercise 1 (page 7)

1 (a) $F(x) = 0$ in $]-\infty, 0[$, $\frac{1}{16}x^2$ in $[0,4]$,
   1 in $]4, \infty[$
   (b) 2.83

2 (a) $F(x) = 0$ in $]-\infty, 0[$, $\frac{2}{3}x$ in $[0,1[$,
   $\frac{1}{3}(4x - x^2 - 1)$ in $[1,2]$, 1 in $]2, \infty[$
   (b) $\frac{3}{8}$ (c) 1.33

3 (a) $F(x) = 0$ in $]-\infty, 0[$, $\frac{3}{28}x^2$ in $[0,2[$,
   $\frac{1}{28}(6x^2 - x^3 - 4)$ in $[2,4]$, 1 in $]4, \infty[$
   (b) 2.17   (c) $\frac{41}{224}$

4 (a) $F(x) = 0$ in $]-\infty, 0[$,
   $\frac{1}{16}(24x^2 - 16x^3 + 3x^4)$ in $[0,2]$,
   1 in $]2, \infty[$
   (b) 7710

5 $f(x) = \frac{1}{6}x - \frac{1}{36}x^3$ in $[0,6]$, 0 otherwise

6 $f(x) = \frac{1}{9}x$ in $[0,3[$, $\frac{2}{3} - \frac{1}{9}x$ in $[3,6]$,
   0 otherwise

7 (a) 0 in $]-\infty, 0]$, $0.2x$ in $]0,5[$, 1 in $[5, \infty[$
   (b) 0 in $]-\infty, 0]$, $0.2\sqrt{y}$ in $]0,25[$,
   1 in $[25, \infty[$
   (c) 0 in $]-\infty, 0]$, $\frac{0.1}{\sqrt{y}}$ in $]0,25[$, 1 in $[25, \infty[$

8 (a) 0 in $]-\infty, 0]$, x in $]0,1[$, 1 in $[1, \infty[$
   (b) 0 in $]-\infty, 1]$, $1 - \frac{1}{y}$ in $]1, \infty[$
   (c) 0 in $]-\infty, 1]$, $\frac{1}{y^2}$ in $]1, \infty[$

9 10.5 years

10 0, 0.216, 0.648, 0.936, 1

11 Data for graph:
   $F(x) = 0$ in $]-\infty, 0[$, 0.183 in $[0,1[$,
   0.493 in $[1,2[$, 0.757 in $[2,3[$, 0.907 in $[3,4[$,
   0.970 in $[4,5[$, 0.992 in $[5,6[$, 0.998 in $[6,7[$,
   1.000 in $[7,8[$, ...

2 Geometric and exponential probability

Exercise 2A (page 16)

1 (a) 0.0720   (b) 0.0504   (c) 0.1577
2 (a) 0.144   (b) 0.0864   (c) 0.4704
3 (a) 0.032   (b) 0.992   (c) 0.0003
4 (a) 0.0199   (b) 0.7763   (c) 0.2237
5 (a) 0.3206   (b) 0.1673   (c) 0.1673
   (d) 0.8327

6 (a) 0.0808   (b) 0.1594   (c) 0.8406
7 (a) 0.3056   (b) 0.0711   (c) 0.0541
8 (a) 0.0453   (b) 0.2711   (c) 0.5811
9 (a) 0.0478   (b) 0.3487
10 (a) 0.0988   (b) 0.1317   (c) 0.1646
   If the people arrive in groups, their choices of
   footwear may not be independent of each other.

11 (a) 0.3894   (b) 66
12 (a) 0.2288   (b) 88
13 (a) 0.0558   (b) 0.2791
14 0.34
15 4, 12
16 (a) 2, 2   (b) 1
17 \$37.50

Exercise 2B (page 23)

1 (a) 0.181   (b) 0.135   (c) 0.148
2 0.713
3 0.154
4 0.223
5 0.435
6 0.221
7 0.082
8 (a) (i) $\frac{1}{\lambda}\ln 2$ (ii) $\frac{1}{\lambda}\ln\frac{4}{3}$, $\frac{1}{\lambda}\ln 4$ (iii) 0
   (b) 63.2%
9 0.383
10 17.9
11 0.865
12 0.916, 0.511
13 1.2 hours

14 (a) 6 minutes   (b) 0.0357

3 Linear combinations of random variables

Exercise 3 (page 33)

1 5, 3.2

2 $18 500, $3000

3 $S = \frac{5}{4}R - 20$, 130

4 0.5, 0.25; 0.48, 0.16; 0.3, 0.37, 0.2, 0.04;
1, 0.5, 0.8, 0.48, 1.8, 0.98

5 3.1, 3.69; 0.2, 0.16, 0.1, 0.14, 0.2

6 24, 12, 6, 24

7 8 mm, 0.14 mm^2

8 3.5, $\frac{35}{12}$, 3, 9, 6.5, $\frac{143}{12}$

9 $\frac{17}{8}$, $\frac{71}{64}$; $\frac{51}{8}$, $\frac{213}{64}$

10 82.5, 311

4 Some properties of normal probability

Exercise 4A (page 40)

1 0.0592

2 0.0636

3 0.0228

4 0.127

5 0.362

6 0.0317

7 $\mu = 16$, $\sigma^2 = 9$

Exercise 4B (page 43)

1 (a) 0.460   (b) (i) 0.382   (ii) 0.159

2 (a) N(2.4, 0.0432) (b) 0.3152

3 (a) 0.271   (b) 0.0021

4 1.3×10^{-11}; 0.631

5 (a) 1.2×10^{-21} (b) 0.9987

6 24.872

7 2.11 hours

5 Hypothesis testing

Exercise 5A (page 47)

1 $H_0: \mu = 102.5$, $H_1: \mu < 102.5$

2 $H_0: \mu = 84$, $H_1: \mu \neq 84$

3 $H_0: \mu = 53$, $H_1: \mu > 53$

4 $H_0: \mu = 30$, $H_1: \mu < 30$. A one-tail test is more appropriate for the customer.

5 $H_0: \mu = 5$, $H_1: \mu < 5$. A one-tail test for a decrease is appropriate since the only cause for concern is μ falling below 5.

Exercise 5B (page 51)

The end-points of the critical regions are given correct to 2 decimal places.

1 (a) Reject H_0 and accept that $\mu > 10$.
(b) Accept $H_0: \mu = 10$.

2 (a) Accept $H_0: \mu = 15$.
(b) Reject H_0 and accept that $\mu < 15$.

3 (a) Accept $H_0: \mu = 20$.
(b), (c) Reject H_0 and accept that $\mu \neq 20$.

4 $H_0: \mu = 102.5$, $H_1: \mu < 102.5$; $\overline{X} \leq 101.69$; accept $\mu < 102.5$.

5 $H_0: \mu = 84.0$, $H_1: \mu \neq 84.0$; $\overline{X} \leq 82.11$ or $\overline{X} \geq 85.89$; reject H_0 and accept that mean time differs from 84.0 s.

6 $\overline{x} = 4.87$; $H_0: \mu = 5$, $H_1: \mu < 5$; $\overline{X} \leq 4.71$; accept H_0, mean is at least 5 microfarads.

7 $H_0: \mu = 85.6$, $H_1: \mu < 85.6$; $\overline{X} \leq 78.39$; reject H_0 and accept that the mean is less than 85.6.

8 $H_0: \mu = 6.8$, $H_1: \mu \neq 6.8$; $\overline{X} \leq 6.69$, $\overline{X} \geq 6.91$; reject H_0, and accept that the mean is not equal to 6.8.

9 $H_0: \mu = 30$, $H_1: \mu < 30$; $\overline{X} \leq 28.94$; accept H_1, there is cause for complaint.

10 $H_0: \mu = 3.21$, $H_1: \mu \neq 3.21$; $\overline{X} \leq 2.92$, $\overline{X} \geq 3.50$; accept H_0; the sample does not differ significantly from a random sample drawn from the population of all of the hospital births that year.

Exercise 5C (page 54)

1 $H_0: \mu = 330$, $H_1: \mu > 330$; $z = 2.504 > 1.960$, so accept H_1, manager's suspicion is justified.

2 $H_0: \mu = 508$, $H_1: \mu \neq 508$; $z = -1.622$. This lies between -1.645 and 1.645, so accept H_0, the process is under control.

3 $H_0: \mu = 45.1$, $H_1: \mu \neq 45.1$; $z = -2.758$. This is outside -2.054 to 2.054, so accept H_1, the mean has changed.

4 $H_0: \mu = 160$, $H_1: \mu > 160$; $z = 2.546 > 2.326$, so accept H_1, that the mean is greater than 160.

5 $H_0: \mu = 276.4$, $H_1: \mu > 276.4$; $z = 1.328 < 1.645$, so accept H_0, the campaign was not successful. This assumes that the sample can be treated as random; normal distribution of daily sales; the standard deviation remains unchanged.

Exercise 5D (page 57)

1 (a) 0.076 (b) (i) No   (ii) Yes

2 $H_0: \mu = 8.42$, $H_1: \mu > 8.42$; $z = 1.703$,

$P\left(\overline{X} > 8.63\right) = 0.044 = 4.4\%$.

(a) Accept H_1, mean greater than 8.42.

(b) Accept H_0, mean not greater than 8.42.

3 $H_0: \mu = 4.3$; $H_1: \mu < 4.3$;
$P\left(Z < -1.581\right) = 0.0569 = 5.69\%$, so accept H_1
that the mean waiting time has decreased.

4 $H_0: \mu = 42.3$, $H_1: \mu > 42.3$;
$P\left(Z > 2.594\right) = 0.00475 = 0.475\%$, so accept
H_1, the results are unusually good.

6 Large sample distributions

Exercise 6A (page 64)

1 0.264

2 0.975

3 $0.961 > 0.95$

4 0.106

5 $N\left(1000, 10000\right)$, 0.159

Exercise 6B (page 66)

1 (a) 0.6928   (b) 0.704

2 (a) 0.0664   (b) 0.641

3 0.108

4 0.0969

5 8.37

Exercise 6C (page 70)

1 (a) 0.0289 (b) 0.114 (c) 0.0806 (d) 0.374

2 (a) 0.0122   (b) 0.117

3 0.0264, 26

4 0.198

5 0.773

7 The central limit theorem

Exercise 7 (page 75)

Where appropriate, alternative answers are given; the
second, shown by [*...], includes a continuity
correction.

1 $N\left(50, 25\right)$
(a) 0.841   (b) 0.977   (c) 0.819

2 (a) 2.8   (b) 0.76   (c) 0.0524 [*0.0443]
(d) 0.895 [*0.911]

3 $P(X = x) = \frac{1}{6}$, $x = 1, 2, 3, 4, 5, 6$
(a) 0.164 [*0.155]   (b) 0.0708 [*0.0662]

4 (a) 0.966 [*0.968]   (b) 0.952 [*0.955]

5 (a) 0.007 86 [*0.008 41]   (b) 81 [*82]

6 0.0101 [*0.0100]

8 Hypothesis testing with discrete variables

Exercise 8A (page 79)

1 $H_0: p = \frac{1}{2}$, $H_1: p \neq \frac{1}{2}$, p is proportion of girls;
$X \sim B\left(18, \frac{1}{2}\right)$; $P\left(X \leq 6\right) = 0.1189 > 0.05$, so
accept H_0, the numbers of boys and girls are
equal.

2 $H_0: p = 0.95$, $H_1: p < 0.95$; $X \sim B(25, 0.95)$;
$P\left(X \leq 22\right) = 0.1271 > 0.05$, so accept H_0, there
are at least 95% satisfied customers.

3 $H_0: p = \frac{1}{2}$, $H_1: p > \frac{1}{2}$ (or $\mu = 2.5$ and $\mu > 2.5$);
p is the proportion of nails with length greater
than 2.5 cm; $X \sim B\left(16, \frac{1}{2}\right)$;
$P\left(X \geq 13\right) = 0.0106 < 0.025$, so accept H_1,
$p > \frac{1}{2}$ and $\mu > 2.5$.
The symmetry of the normal distribution about
its mean is used in the statement $H_0: p = \frac{1}{2}$.

4 (a) $H_0: p = \frac{1}{4}$, $H_1: p > \frac{1}{4}$; $X \sim B\left(20, \frac{1}{4}\right)$;
$P\left(X \geq 8\right) = 0.1018 > 0.05$, so accept H_0,
the results are no better than those obtained
by chance.

(b) 11

5 $H_0: p = \frac{1}{6}$, $H_1: p > \frac{1}{6}$; $X \sim B\left(30, \frac{1}{6}\right)$;
$P\left(X \geq 10\right) = 0.0197 < 0.05$, so accept H_1, the
suspicion is confirmed.

6 (a) $H_0: p = 0.8$, $H_1: p \neq 0.8$; $X \sim B(12, 0.8)$;
$P\left(X \leq 6\right) = 0.0194 < 0.05$, so reject H_0
and accept that the true figure is not 80%.

(b) Lisa's friends do not comprise a random
sample, so test is unreliable.

Exercise 8B (page 81)

1 Between 17 and 31.

2 124 or more

3 151

4 21

9 Errors in hypothesis testing

Exercise 9A (page 87)

1 (a) $\overline{X} \geq 6.316$ (b) 0.196

2 (a) $\overline{X} \leq 9.7419$ (b) Type I error
   (c) 0.141

3 (a) 9.78   (b) 0.204   (c) It will be smaller.

4 (a) $H_0:\mu = 1.94$, $H_1:\mu < 1.94$; rejection region:
       $\overline{X} \leq 1.883$; $\bar{x} = 1.7$, reject H_0 and accept
       that new system had the desired effect.
   (b) 0.00866
   (c) Accepting that the mean absence rate is 1.94
       when it is actually less than 1.94.

5 (a) $\overline{T} \leq 3.3065$ (b) 0.0877
   (c) 0.0646, which is less than 0.0877

6 (a) $H_0:\mu = 30$, $H_1:\mu < 30$; $\overline{X} > 29.7782$
   (b) $0 < \mu < 29.53$

Exercise 9B (page 93)

1 (a) $N \leq 10$, $N \geq 23$ (b) 0.0259
   (c) Accept H_0: $p = 0.55$ (d) 0.9536

2 (a) $R \geq 18$, Mrs Robinson should be elected.
   (b) 2.16% (c) 0.694

3 (a) 0.0635 (b) 0.859;
       α is bigger and β is smaller.

4 (a) $H_0:p = 0.1$, $H_1:p < 0.1$
   (b) $P(X = 0) = 0.0798 < 0.10$, so reject H_0 and
       accept that the modified toaster is more reliable.
   (c) 0.0798   (d) $p < 0.0119$

10 The t-distribution

Exercise 10A (page 101)

1 £1.83, 0.119 £^{2}

2 (a) 48, 0.24   (b) 48, 0.245

3 (a) 0.0138 cm^2 (b) 0.0145 cm^2

4 2.49, 1.61

5 5.6, 2.569

6 50.6 g, 147 g^2

Exercise 10B (page 107)

1 $H_0:\mu = 6.32$, $H_1:\mu \neq 6.32$;
   $|t| = |-0.220| < 2.145$, or $p = 82.9\% > 5\%$;
   accept H_0, mean pH does not differ.

2 $H_0:\mu = 4.8$, $H_1:\mu < 4.8$;
   $t = -3.643 < -2.365$, or $p = 0.41\% < 2.5\%$;
   reject H_0 and accept that adjustment has reduced
   time.

3 $H_0:\mu = 5$, $H_1:\mu > 5$;
   $t = 2.414 > 1.711$, or $p = 1.19\% < 5\%$;
   reject H_0, glass is acceptable.

4 $H_0:\mu = 5.7$, $H_1:\mu < 5.7$;
   $t = -2.994 > -3.365$, or $p = 1.52\% > 1\%$;
   accept H_0, reorganisation has not cut delay.

5 $H_0:\mu = 14.2$, $H_1:\mu \neq 14.2$;
   $|t| = |1.410| < 2.262$, or $p = 19.2\% > 5\%$;
   accept H_0, level for students does not differ.

6 $\left|\dfrac{3.207 - \mu}{0.461/\sqrt{20}}\right| < 1.729$; $3.029 < \mu < 3.385$

Exercise 10C (page 109)

1 $H_0:\mu_1 = \mu_2$, $H_1:\mu_1 > \mu_2$;
   $t = 1.279 < 1.895$, or $p = 12.1\% > 5\%$;
   accept H_0, the belief is not suppported.

2 $H_0:\mu_D = 40$, $H_1:\mu_D > 40$;
   $t = 2.547 > 2.015$, or $p = 2.57\% < 5\%$;
   reject H_0 and accept that the difference is more
   than 40.

3 $H_0:\mu_D = 10$, $H_1:\mu_D > 10$;
   $t = 3.162 > 2.821$, or $p = 0.58\% < 1\%$;
   reject H_0 and accept that time goes up by more
   than 0.1 s.

4 $H_0:\mu_o = \mu_c$, $H_1:\mu_o \neq \mu_c$;
   $|t| = |-0.661| < 2.228$, or $p = 52.4\% > 5\%$;
   accept H_0, the mean yields are the same.

5 $H_0:\mu_D = 0$, $H_1:\mu_D > 0$;
   $t = 1.554$, or $p = 6.41\%$;
   accept H_0 at the 5% level, the pill makes no
   difference.

11 Confidence intervals

Exercise 11A (page 118)

1 $[496.5, 504.5]$, 190

2 (a) $[0.981, 1.028]$ (b) 0.0233 litres

3 (a) 5.155   (b) 8.48   (c) 10.10

4 $[169.4, 181.0]$; assumes a random sample;
   no, 178 is within the interval.

5 (a) $[5.013, 5.057]$
   (b) 5.00 cm is outside the interval; this indicates
       that the mean has increased.

6 (a) $[15.67, 16.17]$ (b) Six more

7 $[65.1, 70.5]$; this includes 70, so the test is
   satisfactory and can be used unchanged.

Exercise 11B (page 121)

1 (a) 1.796 (b) 2.262 (c) 2.492
   (d) 2.871

2 $[14.66, 15.20]$; assumes alcohol content has a normal distribution.

3 $[2725.8, 2967.4]$; 10% chance that interval will not contain population mean.

4 (a) $[3.65, 4.75]$; assumes a normal population and a random sample.
   (b) Class appears better than average, suggesting that they are not a random sample.

5 (a) 9.806 (b) $[9.756, 9.856]$
   (c) Interval is exact; neither the distribution nor parameters are approximated.

6 (a) $[18.3, 19.9]$
   (b) 15.9 is outside 98% interval. The swallow is most unlikely to be fully grown.

7 (a) 498.1 g, 4.839 g^2 (b) $[497.7, 498.5]$
   The confidence interval will not contain the mean 5% of the time.

8 (a) 2.43, 2.288 (b) $[2.18, 2.68]$

Exercise 11C (page 125)

1 $[28.5\%, 39.5\%]$

2 (a) $[0.144, 0.224]$ (b) Sample is random.
   (c) For example, all cars passing under the bridge over a specified period
   (d) 49

3 $[0.322, 0.424]$; $n = 971$

4 $[0.640, 0.860]$, $n = 180$

5 (a) Confidence interval for p is $[0.427, 0.560]$; classify as average.
   (b) 95% confidence interval $[0.413, 0.573]$; yes

12 A survey of probability distributions

Exercise 12 (page 134)

2 (a) 3, 2   (b) 0.4, 0.24
   (c) 4, 2.4   (d) 2.5, 3.75
   (e) 7.5, 11.25 (f) 1.2, 0.56
   (g) 2.4, 2.4

5 (a) 0.258 (b) 6.72 (c) 0.042 (d) 0.0280

6 (a) 0.127 (b) 0.323 (c) 0.0593 (d) 4
   (e) 0.0645 (f) 40, 3.65

7 (a) 0.234 (b) 4.8, 1.49

8 (a) $\frac{1001}{4845}, \frac{2184}{4845}, \frac{1365}{4845}, \frac{280}{4845}, \frac{15}{4845}$
   (b) 1.2, 0.707 (c) $\frac{6}{5}, \frac{336}{475}$

9 Means are both 6; variances are 4.2, 3.39.

10 (a) 0.287 (b) 0.123 (c) 0.265

11 100 s, 31.6 s

12 (a) 0.593 (b) 0.224

13 Chi-squared tests

In all these exercises, the answers are presented in the format: distribution under H_0; expected frequencies (after combination of classes if necessary); value of X^2; $P(\chi_v^2 \geq \text{value of } X^2)$ as a percentage; accept or reject H_0. Your estimated frequencies and values of χ^2 may differ from those given, depending on how much estimated parameters are rounded. This should not affect your conclusions.

Exercise 13A (page 149)

1 All $\frac{1}{6}$; all 10; 3.4; 63.9%; accept H_0.

2 $p_1 = \frac{9}{16}$, $p_2 = p_3 = \frac{3}{16}$, $p_4 = \frac{1}{16}$; 90, 30, 30, 10; 1.84; 60.6%; accept H_0.

3 (a) $B\left(4, \frac{1}{4}\right)$; 63.28, 84.37, 42.19, 10.16; 241.85; negligible; reject H_0. (3, 4 combined)
   (b) $B(4, 0.465)$; 16.38, 56.96, 74.27, 43.03, 9.35; 3.69; 29.7%; accept H_0.

4 $B\left(6, \frac{5}{18}\right)$; 8.51, 19.65, 18.89, 12.95; 2.34; 31.0%; accept H_0. (3, 4, 5, 6 combined)

5 $Geo\left(\frac{1}{2}\right)$; 25, 12.5, 6.25, 6.25; 2.24; 52.4%; accept H_0. (4, 5, 6, 7 or more combined)

6 All $\frac{1}{10}$; all 9; 16; 6.69%; reject H_0.

7 $Po\left(\frac{13}{15}\right)$; 37.83, 32.79, 14.21, 5.17; 4.91; 8.59%; accept H_0. (3, 4 or more combined)

8 All $\frac{1}{7}$; all 85.71; 3.10; 79.6%; accept H_0.

9 $p_1 = \frac{9}{49}$, remainder $\frac{10}{49}$; 110.2, remainder 122.45; 3.04; 55.1%; accept H_0.

10 $B(5, 0.424)$; 14.84, 17.18, 12.64, 5.34; 2.47; 29.1%; accept H_0. (0, 1 and 4, 5 combined)

11 (a) $Po(1)$; 30.90, 30.90, 15.45, 6.75; 26.84; 0.000 64%,; reject H_0. (3, 4, 5, 6, 7 or more combined)
   (b) $Po\left(\frac{67}{42}\right)$; 17.04, 27.18, 21.68, 11.53, 6.57; 1.81; 61.3%; accept H_0. (4, 5, 6, 7 or more combined)

12 (a) 1.3, 0.769 to 3 sig.fig.
   (b) $Geo(0.769)$; 76.92, 17.75, 5.33; 0.566; 45.2%; accept H_0. (3, 4, >4 combined)

13 $Po\left(\frac{45}{52}\right)$; 21.89, 18.94, 8.20 + 2.97 = 11.17; 4.03; 4.47%; accept H_0. (2, 3, 4, 5 or more combined)

Exercise 13B (page 154)

1 $N(50,10^2)$; $2.28+13.59=15.87$, 34.13, 34.13, $13.59+2.28=15.87$; 0.89; 82.8%; accept H_0.

2 (b) $N(100,15^2)$; 6.83, 40.77, 102.40, 102.40, 40.77, 6.83; 17.61; 0.35%; reject H_0.
   (c) $N(96.6,15^2)$; 11.43, 54.47, 110.99, 90.11, 33.00; 3.04; 38.5%; accept H_0.

3 (b) $N(698.3,18.83^2)$; 7.95, 8.05, 10.26, 9.93, 7.29, 6.51; 3.36; 33.9%; accept H_0.

4 (c) 8.70, 24.99, 40.03, 26.27; 7.34; 6.2%; reject H_0.

5 $N(172.2,7.31^2)$; 8.98, 11.42, 13.31, 9.89, 6.40; 0.17; 91.9%; accept H_0.

6 $N(2.915,1.775^2)$; 14.03, 16.28, 21.60, 21.04, 15.04, 12.01; 18.60; 0.033%; reject H_0.

Exercise 13C (page 161)

1 9, 24, 27, 36, 96, 108, 30, 80, 90; 18.97; 0.08%; reject H_0.

2 32.5, 68.25, 29.25, 17.5, 36.75, 15.75; 2.44; 29.5%; accept H_0.

3 32.64, 20.40, 14.96, 15.36, 9.60, 7.04; 2.62; 27.0%; accept H_0.

4 71.5, 53.6, 69.9, 85.1, 63.8, 83.1, 48.8, 36.6, 47.7, 14.7, 11.0, 14.3; 29.0; 0.0061%; reject H_0.

5 101.31, 82.28, 123.41, 149.16, 121.14, 181.70, 79.53, 64.59, 96.88; 44.28; $5.6\times10^{-7}\%$; reject H_0.

6 (a) 14.79, 35.09, 17.54, 9.29, 4.47, 3.44, 1.38, 12.90, 30.60, 15.30, 8.10, 3.90, 3.00, 1.20, 15.31, 36.31, 18.16, 9.61, 4.63, 3.56, 1.42.
   (b) 9 of the expected frequencies are less than 5.
   (c) H_0: 14.79, 35.09, 17.54, 9.29, 9.29, 12.90, 30.60, 15.30, 8.10, 8.10, 15.31, 36.31, 18.16, 9.61, 9.61; 35.148; 0.0025%; reject H_0. (E, N and U combined).

7 (a) 16.15, 13.77, 4.08, 44.65, 38.07, 11.28, 34.20, 29.16, 8.64
   (b) One of the expected frequencies is less than 5.
   (c) H_0: no association; 60.80, 51.84, 15.36, 34.20, 29.16, 8.64; 5.79; 5.5%; accept H_0 (first two rows combined).

Review exercise

(page 164)

1 (a) 0.1853 (b) 0.1839

2 (a) $\frac{9}{7}$, $\frac{3}{49}$
   (b) 0.0289; distribution of T is very skewed so answer has considerable error.

3 (a) 0.0593 (b) 0.0023

4 (a) 2.7, $\dfrac{2.7}{n}$ (b) 0.0192

5 (a) H_0: $\mu=2.855$, H_1: $\mu \neq 2.855$; $Z \leq -1.96$, $Z \geq 1.96$.
   (b) $z=-1.540$, so accept H_0: $\mu=2.855$, the batch is from the specified population.

6 H_0: $p=0.07$, H_1: $p>0.07$; $X \sim B(125,0.07)$; $P(X \geq 14)=0.551>0.03$, accept H_0 and retain the batch.

7 (a) H_0: $p=0.1$, H_1: $p>0.1$,
   (b) $r>4$; for $r=4$ do not reject H_0.
   (c) 0.6296

8 (a) 2
   (b) $f(v) = \begin{cases} \dfrac{90}{v^2} & \text{for } 22.5 \leq v \leq 30, \\ 0 & \text{otherwise.} \end{cases}$
   (c) (i) $25\frac{5}{7}$ km per hour
        (ii) $90\ln\frac{4}{3}$ km per hour

9 (a) 0.682 (b) 0.113 (c) 0.159

10 (a) $t=-2.515 \notin [-1.753,1.753]$, or $P(|T|>2.515)=2.38\%<10\%$; reject H_0 and accept that the mean is not 4 kg.
    (b) $\mu<1.97$, $\mu>3.83$

11 (a) Assumes students form a random sample and T_1-T_2 has a normal distribution; H_0: $\mu_1-\mu_2=0$, H_1: $\mu_1-\mu_2>0$; $t=2.504>1.833$, or $P(T>2.504)=1.68\%<5\%$, so reject H_0 and accept that there is a reduction in the mean time.
    (b) $[0.66, 12.92]$, to 2 decimal places

12 $Po(2)$; 16.24, 32.48, 32.48, 21.65, 10.83, 6.32; $P(\chi_4^2 \geq 12.24)=1.57\%$; reject H_0 at 5% significance level (5, 6, 7 combined).

13 H_0: age and sex are independent; 202.2, 260.7, 318.1, 184.8, 238.3, 290.9; $P(\chi_2^2 \geq 2.022)=36.4\%$; accept H_0.

Examination questions

1 0.164

2 $[2.7030, 2.7070]$

3 $P(\chi_3^2 \geq 4.907)=17.9\%$; accept that sons follow fathers.

4 (a) (i) 87.13 km h^{-1} (ii) 215.58 $\left(\text{km h}^{-1}\right)^2$

(b) (i) $[86.22, 88.04]$ (ii) $[86.37, 87.89]$

(c) 95% probability requires a wider interval than 90%.

5 (a) (i) 1.98   (ii) 0.33

(b) $P\left(\chi_3^2 \geq 1.563\right) = 66.8\%$; a good fit.

6 (a) 0.0784, 0.216, 0.296, 0.28, 0.1296

(b) $P\left(\chi_3^2 \geq 4.825\right) = 18.5\%$; accept assumption.

7 (a) 10   (b) 12   (c) 7   (d) 35

8 $P\left(|Z| \geq 2.020\right) = 0.0434$; accept H_0, the mean length is 1.005 m.

Appendix

(page 169)

2.1 $1 + q + q^2 + q^3 + \dots$. This is the infinite geometric series with sum $\dfrac{1}{1-q} = (1-q)^{-1}$, where $-1 < q < 1$.

4.1 $-xe^{-\frac{1}{2}x^2}$

4.2 $\left(\pm 1, \dfrac{1}{\sqrt{2\pi e}} \right)$

4.4 The standard deviation is represented by the distance from a point of inflexion to the y-axis.

5.3 $\mu = \displaystyle\sum_{x=0}^{n} xp_x$, which can also be written as

$\displaystyle\sum_{x=1}^{n} xp_x$.

6.1 $ve^{-mv} - \dfrac{1}{m}e^{-mv} + \dfrac{1}{m}$

Index

The page numbers refer to the first mention of each term, or the box if there is one.